Plastics Waste Management

Scrivener Publishing
100 Cummings Center, Suite 541J
Beverly, MA 01915-6106

Publishers at Scrivener
Martin Scrivener (martin@scrivenerpublishing.com)
Phillip Carmical (pcarmical@scrivenerpublishing.com)

Plastics Waste Management

Second Edition

Muralisrinivasan Natamai Subramanian

Scrivener Publishing

WILEY

This edition published 2019 by John Wiley & Sons, Inc., 111 River Street, Hoboken, NJ 07030, USA and Scrivener Publishing LLC, 100 Cummings Center, Suite 541J, Beverly, MA 01915, USA

For more information about Scrivener publications please visit www.scrivenerpublishing.com.

1st edition (2016), 2nd edition (2019)

Wiley Global Headquarters
111 River Street, Hoboken, NJ 07030, USA

For details of our global editorial offices, customer services, and more information about Wiley products visit us at www.wiley.com.

Library of Congress Cataloging-in-Publication Data

Names: Muralisrinivasan, Natamai Subramanian, author.
Title: Plastics waste management : processing and disposal /
Muralisrinivasan Natamai Subramanian.
Description: Second edition. | Hoboken : Wiley, 2019. | Includes bibliographical references and index.
Identifiers: LCCN 2019026450 (print) | LCCN 2019026451 (ebook) | ISBN 9781119555872 (hardback) | ISBN 9781119556190 (adobe pdf) | ISBN 9781119556183 (epub)
Subjects: LCSH: Plastic scrap. | Factory and trade waste. | Plastics--Environmental aspects.
Classification: LCC TD798 .M87 2019 (print) | LCC TD798 (ebook) | DDC 628.4/4--dc23
LC record available at https://lccn.loc.gov/2019026450
LC ebook record available at https://lccn.loc.gov/2019026451

Cover image: Pixabay.Com
Cover design by Russell Richardson

Set in size of 12pt and Minion Pro by Manila Typesetting Company, Makati, Philippines

10 9 8 7 6 5 4 3 2 1

Contents

Preface

Plastics are increasingly being used in modern-day life due to their wide-ranging properties which can be used in a variety of applications. The importance of plastics is reflected in this second edition of *Plastics Waste Management: Processing and Disposal*, which essentially evolved from the well-regarded previous edition and includes updated information on plastics waste management, from generation to collection, processing and final use or disposal. Primarily written from an industrial point of view, the book gives readers interested in waste management a better understanding of the correct processes for dealing with the growing environmental problem of plastics waste from the perspective of utilization and reduction. Since plastics waste management can be promoted through improvements, it is hoped that the fresh ideas presented in this book will encourage developed countries to further implement the methods suggested; that developing countries will become more involved in the process; and that underdeveloped countries will start becoming involved by studying and implementing the waste management programs based on the concepts and fundamentals presented.

With years of experience in various aspects of the processing field as well as planning, the contributing authors of the book delve into all aspects of plastics waste processing and disposal, including non-technical considerations, to produce a comprehensive approach to plastics waste management. Although the book mainly focuses on plastics waste, plastic materials, additives and processing are dealt with to some extent and thermosetting materials are also briefly covered. In addition, economic aspects are considered along with technical details.

The primary objective of this book is to serve as a platform for the exchange of in-depth information based on logical reasoning.

It would only be possible to cover all the plastics and additives in detail in a volume of this size, and this book does not disappoint with the wealth of information contained herein.

This second edition includes various aspects of environmental and economic issues, as well as technological ones. Since some of the previously covered areas have seen so much growth and rapid development since the publication of the first edition, new chapters that address plastics waste utilization and their processes have been added. The chapters from the first volume have been reorganized with the major revision being a separation of the overview of the development in plastics waste research into its own expanded chapter covering not only materials but also plastics waste materials processing and its evolving utility on a commercial scale. There is a section on plastics processing management which describes the methods of manufacturing plastic products and a chapter dealing with case studies which gives practical guidance as well as considerable information regarding recycling and disposal. Chapters from the first edition have been expanded to include new references and many of the detailed examples and illustrations from the first edition are also included. All of these chapters continue to provide a clear and concise description of the development in plastics waste processing.

This book examines plastics waste from the perspective of utilization and reduction. It provides clear explanations for newcomers to the subject as well as contemporary details and theory for the experienced user in plastics waste management. It is designed to accompany management or technology courses in business schools and universities, as well as being relevant to academic research departments.

In conclusion, I would like to thank the small army of support that helped to write this book, my wife Mrs. Himachalaganga, and the encouragement of my teachers to get the job done. Special thanks to Mr. Martin Scrivener and others who helped me complete this book. Above all, my sincere gratitude goes to the almighty Lord Nataraja and my teacher Lord Murugan.

Muralisrinivasan Natamai Subramanian
Madurai, India
July 2019

1
Introduction

Within a specific ecological–economic system, each material is generally connected to and also dependent on others. Moreover, plastics contribute to the overall integrity of this life system. However, since all plastics contribute to the functional scheme of things in waste generation, it is undoubtedly a very important issue. In particular, plastics provide key applications that are also unique; without plastics, the underlying ecological–economic system would be very different.

The first synthetic plastics developed were celluloid (cellulose nitrate) in 1869 and phenol formaldehyde in 1909. Other plastics such as cellulose acetate and polyvinylchloride were made into semi-durable items, such as electrical equipment or insulation, motion picture film, billiard balls, etc., which are considered a nuisance when they become waste. Plastics production and consumption have increased considerably since the first industrial production of plastics in the 1940s [1]. The high-volume production of low-density polyethylene began in 1940, which is also when plastic waste started being recycled. Included in the rapidly growing plastics industry are all thermoset plastics and thermoplastics, with the consumption of plastic materials having grown to around one million tons per year as of 1962. Worldwide production and consumption of plastics has increased at an average rate of about 8 percent per annum [2]. However, it is now a billion tons of plastics waste. Plastics consumption has increased rapidly while the deposit of natural resources is decreasing. The decrease in crude oil and natural gas puts pressure on plastics production. Therefore, rapid, large changes in oil prices can cause significant long- and short-term economic consequences. Obtaining and using this oil also carries with it the enormous burden of adverse environmental consequences, social

Muralisrinivasan Natamai Subramanian (ed.) Plastics Waste Management, 2nd Edition (1–6)

issues, and geopolitical risk, since plastics undergo little degradation and dispersion by natural processes.

Global post-consumer waste generation totals approximately 900–1,250 metric tons per year [3, 4]. In an underdeveloped country, the per capita solid waste generation rate is less than 0.1 tons per capita per year as opposed to developed countries where it is greater than 0.8 tons per capita per year in high-income industrialized countries [5–12].

Plastics waste is closely linked to population type and size, and the degree of urbanization and material comfort. It remains a major challenge for municipalities to collect, recycle, treat and dispose of increasing quantities of plastics waste in most developed and developing countries. Most technologies for plastics waste management are immature and have been difficult to implement in many countries.

Plastic waste has gone up both in absolute terms and as a percentage of solid waste. However, the volume may not be enough to warrant systems to separate different types of plastics from each other for recovery. Because of the amount of plastic waste disposed of in municipal solid waste, it needs to be managed.

Plastics waste can be used as a raw material for recycling operations or can be treated prior to disposal, resulting in the waste being transformed into material which can be safely disposed of or reused. The proper management of plastics waste starts at the production stage. Plastics waste has an economic advantage, in comparison with many other solid wastes, as it can be regularly recycled. Current processing technology enables the efficient conversion of waste into new recycled end products.

Plastics waste management does not exist in a vacuum; waste plastics are affected by and impact upon many different aspects of national life, i.e., there is a balance between the utilization of plastics waste and its production and processing. The majority of plastics waste generation is related to material comfort items; however, recycling/reuse initiatives for mixed plastics are limited [3].

In particular, it is crucial that plastics waste management is linked to the parallel development of production and processing, otherwise there is a risk that controls to limit the environmental pollution of one operation will lead to an increased level of pollution in another, hence:

- Plastic processes and activities should be chosen which produce the lowest amount of waste.
- The production of hazardous waste from antimony and lead and so on from additives, should be kept to a minimum.
- All feasible and reasonable steps should be taken to recycle and reuse materials from plastics waste and convert this waste into useful marketable products.
- The waste disposal process should include arrangements for the disposal of plastics waste that cannot be reclaimed, such as degraded polyvinyl chloride. Disposal should reduce the level of risk to public health, water supplies and the environment to acceptable levels.
- All types of solid waste should only be disposed of at sites suitable for the disposal of that particular waste, which will not be reclaimed. The site can stipulate upon acceptance, any special requirements regarding the method of deposal which includes preparation to receive the waste, the methods involved in disposing of the waste and so on.
- Plastics waste treatment and the methods to be used for the disposal of the residues from the treatment should be included in the waste disposal process.
- Waste generators are responsible for their waste, which is a very important aspect for plastics waste. Generators of waste must be assumed to have adequate knowledge of its composition, form, and of the potential hazards to public health and the environment, to ensure disposal of the waste is not detrimental to the environment. The waste generator is responsible for ensuring that only appropriate disposal methods are used for their waste.
- Future planning needs to include the proper management of plastics waste.

Recycling of post-consumer plastics has not yet become a significant recovery option. Plastics pollution in most cases results in already stressed ecosystems. Humans fear that the dangers posed by plastics waste tend to create problems more often than not. An attempt is being made to treat plastics-waste-related environmental and natural resource problems as part of an important task to help the societies of the world. The world has an emerging interest in moving away from plastics waste towards material management due to their non-degradable nature. There are strong drivers at all levels towards a culture of more sustainable plastics waste management.

Industrialists should know the type and quantity of waste produced by their operations and processes, whereas a waste generator should know the composition, properties and environmental impact of the waste. Without this knowledge industrialists cannot properly manage their operations and cannot discharge their responsibilities to protect the health and safety of employees, i.e., the nature of the waste they are exposed to must be known, otherwise they are not in full control of their operation; in addition, if the quantity of waste is unknown the cost, material balance and efficiency cannot be determined [13].

When solid waste including plastics waste disappears from an ecological–economic system, the system changes dramatically. In fact, what is particularly significant is that the disappearance of plastics often triggers the loss of other applications, and when this happens, the complex connections among nexus components, such as packaging with other substitute material, begin to evolve. Minimizing solid waste through an ecological–economic system in effect addresses environmental problems.

References

1. Albano, C., Camacho, N., Hernandez, M., Matheus, A., Gutierrez, A., Influence of content and particle size of waste pet bottles on concrete behaviour at different w/c ratios. *Waste Manage.*, 29, 2707–2716, 2009.

2. Bernardo, C.A., Simões, C.L., Lígia, M., Costa Pinto, Proceedings of the Regional Conference Graz – Polymer Processing Society PPS. *AIP Conf. Proc.*, 1779, 140001-1–140001-5, 2015.
3. Bogner, J. and Matthews, E., Global methane emissions from landfills: New methodology and annual estimates 1980–1996. *Global Biogeochem. Cycles*, 17, 341, 2003.
4. Monni, S., Pipatti, R., Lehtilä, A., Savolainen, I., Syri, S., in: *Global Climate Change Mitigation Scenarios for Solid Waste Management. Technical Research Centre of Finland*, VTT Publications, Espoo, Finland, 2006.
5. Bernache-Perez, G., Sánchez-Colón, S., Garmendia, A.M., Dávila-Villarreal, Sánchez-Salazar, M.E., Solid waste characterisation study in the Guadalajara Metropolitan Zone, Mexico. *Waste Manage. Res.*, 19, 413, 2001.
6. Diaz, L.F. and Eggerth, L.L., *Waste Characterization Study: Ulaanbaatar, Mongolia: Winter-Summer 2002*, Final Report, CalRecovery, Inc. prepared for WHO/WPRO, Manila, Philippines, August, 2002.
7. Kaseva, M.E., Mbuligwe, S.B., Kassenga, G., Recycling inorganic domestic solid wastes: Results from a pilot study in Dar es Salaam City, Tanzania, *Resour. Conserv. Recycl.*, 35, 243, 2002.
8. Idris, A., Bulent, I., Hassan, M.N., Overview of municipal solid waste landfill sites in Malaysia. In: *Proc. of the Second Workshop on Material Cycles and Waste Management in Asia*, Tsukuba, Japan, 2003.
9. Ojeda-Benitez, S. and Beraud-Lozano, J.L., Characterization and quantification of household solid wastes in a Mexican city. *Resour. Conserv. Recycl.*, 39, 211, 2003.
10. *Waste Analysis and Characterization Study, Asian Development Bank, Report TA 3848-PHI*, CalRecovery, Inc., UNEP, Japan, 2004.
11. Griffiths, A.J. and Williams, K.P., Thermal treatment options, Waste Management World, pp. 63–73, July–August 2005.
12. Huang, Q., Wang, Q., Dong, L., Xi, B., Zhou, B., The current situation of solid waste management in China, *J. Mater. Cycles Waste Manage.*, 8, 63, 2006.
13. Rushbrook, P .E. and Finnecy, E.E., Planning for future waste management operations in developing countries, *Waste Manage. Res.*, 6, 1, 1988.

2
Plastics and Additives

Plastic materials are universally used in all societies. The advantages of plastics have contributed decisively to lifestyle improvements. Plastics are ubiquitous materials with the advantageous characteristics of hydrophobic materials. They are used in the domestic market as packaging materials to retard the deterioration of food and to ensure the safety of food at a later date. Their contribution in medical and biomedical applications helps in many applications such as syringes, artificial muscles, optical lenses, etc. As industrial parts, plastics help to increase the lifespan of machine parts due to corrosion prevention and wear in comparison to metal parts, which are by nature non-lubricating and susceptible to oxidation reaction. Therefore, plastics contribute decisively to improve the standard of living of a society [1]. Since plastics are derived from oil and gas, their recycling conserves natural and renewable resources [2].

The global production of plastics, excluding fibers of PET, PP, and polyamide, was approximately 322 million tons in 2015 and 335 million tons in 2016. Around 50% of plastics are made for single-use applications, such as disposable consumer items and packaging; the rest is used for long-term infrastructures, such as pipes and structural materials, and for durable consumer applications with intermediate lifespan such as in furniture and vehicles in almost equal amounts [3].

2.1 Polymers

Polymer formation processes have led to the development of plastics with properties tailored to specific applications. The addition of elements such as carbon, hydrogen, nitrogen, oxygen, fluorine, silicon, sulfur and chlorine, has enabled the production of thousands of

Muralisrinivasan Natamai Subramanian (ed.) Plastics Waste Management, 2nd Edition (7–36) © 2019 Scrivener Publishing LLC

types of plastic materials. Polymers are synthetic materials composed of chemical compounds called monomers, which are chemical units that react together to form long chains of repeating units. Monomers are in a gas or liquid state at the start of the polymer production process. Polymers are produced via polymerization techniques, such as anionic, cationic, step-growth and so on, which result in the formation of large molecules from monomers. The use of one type of monomer results in a homopolymer, whereas the use of two or more different monomers leads to the formation of copolymers. Polymerization is usually controlled by the addition of a catalyst or initiator, and polymers produced by these techniques are thermoplastic in nature.

During the process, monomers join with each other and the chain length grows until the reactive ends combine, which stops chain growth at that point. During the polymerization reaction, the polymer chains simultaneously grow in length. The addition of predetermined inhibitors (chain growth stoppers) can produce polymers with a consistent average chain length, which is a key factor in determining many properties of the plastic and its processing characteristics. It is possible to manufacture polymer molecules with an average molecular weight(s) (MW) distribution and there can be either a broad or narrow spread between the MW of the largest and smallest molecules. A narrow MW distribution provides more uniform properties and a broad distribution material is easier to process. Increasing the chain length increases toughness, creep resistance, stress-crack resistance, melt temperature (Tm), melt viscosity and leads to processing difficulty due to the non-degradable nature of the polymer.

2.2 Plastics

Polymers are ideal materials used in thousands of industrial and consumer products. A major application area of polymers is plastics. The key difference between plastics and polymers is that all plastics are polymers but not all polymers are plastics. A polymer can either be biological or synthetic. The majority of plastics are derived from petroleum or natural gas [4].

Plastics generally consist of macromolecules called polymers and additives which are used to modify the properties of the material;

polymers rarely consist of pure polymer. Additives are used in the production of plastics to enhance the appearance, improve strength and change certain characteristics. Plastics is converted into a finished product via viscous state from solid material, which can be achieved via heat, pressure or a combination of the two.

2.3 Plastics Raw Material

Plastic raw materials are used in the plastics processing industry to produce finished products. They are thermoplastics and thermosets.

1) Thermoplastics are recyclable and used in four segments of the plastics processing industry, which can be distinguished as:
   - Production of plastic plates, film, tubes and molded sections
   - Production of packaging material, more than half of which consists of tins, crates, fittings, balloons, bottles, plugs, lids, etc.
   - Production of construction-related articles, most importantly doors, windows and window frames, wall coverings and window sills.
   - Production of other plastic goods, more than half of the value of this segment consists of technical parts, of which those destined for the automobile industry are the most important. This segment also encompasses household articles made of plastic.
2) Thermosets are not recyclable. The waste after use can be used for incineration to convert energy or aggregates as filler in different applications.

2.4 Thermoplastics

Thermoplastics of recycled types are either commodity plastics or engineering plastics. The commodity plastics include polyethylene (PE), polypropylene (PP), polyvinylchloride (PVC) and polyethylene terephthalate (PET). The cost and physical properties explain their widespread use in disposable packaging. Plastics can be economically produced in large quantities. Engineering plastics are characterized by higher strength, heat resistance, and impact resistance. They have become widely used to replace metals in many automotive, appliance, aerospace and industrial products. These plastics are durable and have low production volume, which results in fewer pollution problems with the exception of PVC and PET. Some of these thermoplastic materials, such as PVC, PET, etc., are accepted in engineering application due to their physical properties.

Thermoplastics soften or melt during heating and regain rigidity when cool. Various processing techniques have been developed for molding a specific shape by exploiting the properties of thermoplastics. No chemical change or crosslinking of molecules occurs during processing. As thermoplastics will soften during reheating, i.e., heat moldable.

2.4.1 Polyolefin

Commodity plastics, mainly polyolefin, are long lasting with high persistence in the environment. Polypropylene belongs to a class called polyolefin. The monomers are olefins made by thermally cracking the large hydrocarbon molecules in natural gas liquids or petroleum fractions. The polyethylene can be thermally cracked or depolymerized back to lower molecular weight polymers or even, under severe conditions, to oily materials or to carbon and hydrogen. This process has been proposed as a method of recycling.

The bulk properties of polyolefin (PO) can be tailored via catalyst design during polymerization process. PO are important polymers which are derived from relatively inexpensive feed stocks and are generally hydrophobic in nature. PO require surface modification to improve certain properties such as adhesion, wettability, printability and biocompatibility. Upon exposure to light or heat, PO usually exhibit an induction period, during which minimal oxygen uptake or changes in physical properties are observed [5]. PO are subject to thermal and oxidative degradation and cannot be used in practical applications, such as automotive parts, unless they are protected with efficient antioxidants [6].

2.4.1.1 Polyethylene

Polyethylene is produced from ethylene monomer with organic peroxide with pressure and temperature. The high pressure and low temperature produces high density polyethylene and high temperature and low pressure produces low density polyethylene. Polyethylene (PE) is one of the most degradation-resistant plastics when disposed of in the environment. Even in and on the soil, it does not display any visible changes and has been used for long-term exposure in the sea for applications such as cable covering and sonar devices. It continues to be the most important plastic in terms of volume used for manufacturing. The low-density polyethylene (LDPE) market continues to dominate with approximately 75% consumption for packaging film manufacturing, followed by extrusion coating and injection molding products; LDPE comprises 64% of the global plastic material manufactured in the form of packaging and bottles, which are usually discarded after only brief use [7].

In theory, PE should be structurally stable against photo-oxidation; however, it may undergo loss of mechanical properties upon outdoor exposure if it includes branching, impurities or residual catalytic species. Plastic bags accumulate in the environment due to their low degradability, hence generating pollution and taking space in landfills. In addition, as they generate very small mass and are usually contaminated, recycling is economically unfeasible [8, 9]. Their elimination at composting plants is not complete, therefore, fragments of bags end up contaminating the compost which ultimately requires screening or other processes to remove it. Table 2.1 illustrates some of the physical properties of PE.

Polyethylene degradation occurs in the life cycle of a product during compounding and molding. It also degrades during product service and recycling or less ideally when buried in a landfill or during combustion. However, except for combustion, degradation occurs most rapidly during melt processing and high mechanical shear between 180°C and 280°C. The available oxygen and its concentration has an important role in thermo-mechanical degradation. The latter degradation is based on the method of polyethylene production [10–13].

High volume usage of polyethylene ensures that there are large quantities of post-consumer HDPE, plastics bags, homopolymers

Table 2.1 Physical properties of PE.

Physical parameters	Value	Units
Specific gravity	0.91–0.94	–
Melting point	110–135	°C
Tensile strength	10–60	Mpa
Elastic modulus	550–1,000	KPa
Hardness	45–70	Shore D

such as milk and juice bottles and copolymers such as shampoo bottles for recycling. The resultant recyclate has essentially the same rheological properties as the virgin resin since it does not undergo any appreciable thermal degradation during recycling [14]. Commodity plastics, e.g., high density polyethylene (HDPE), are versatile, durable, light and cheap, explaining the increasing use of plastic products and the subsequent rise in plastic waste generation.

PE represents the largest constituent of plastics in the municipal waste stream and the fraction is mainly composed of high-density polyethylene (HDPE) packaging materials. Recycling this type of waste packaging material yields a stream of recycled plastic that is highly homogeneous and consistent [15].

2.4.1.2 Polypropylene

Polypropylene (PP) is one of commodity plastics with a high production volume and is used extensively as the plastic material for appliances such as refrigerators, washing machines and air conditioners. Polypropylene belongs to a class called polyolefin. PP has an excellent balance of physical and mechanical properties, and the products based on PP are economical and can be easily recycled. It neither reacts nor interacts with most materials. This characteristic makes it attractive for many packaging applications. The bulk properties of PP are attractive and can be varied via synthesis conditions; however, the surface properties are not easily controllable [16]. Due to its low surface energy, the adhesion of PP to other polar materials is generally low, which significantly limits its engineering applications. To increase the surface energy, surface modification via chemical or physical means is necessary [17].

PP is light-weight, which is an important factor for material substitution in automotive engineering. It allows greater freedom of design due to its impact strength and can replace many conventional materials. PP can withstand higher temperatures than PS and PE, which makes it suitable for use in automotive components [18]. PP is a highly hydrophobic and non-polar polymer and can be used for the preparation of nanocomposites only after a compatibilizing process in which polar groups are introduced [19]. PP belongs to the semi-crystalline PO family and is resistant to chemicals and abrasion; its advantages include higher rigidity over other PO and due to the presence of tertiary hydrogen in the backbone chain, PP is very susceptible to degradation [20].

Table 2.2 illustrates some of the physical properties of PP.

PP is widely used in various applications requiring good impact properties. Using various elastomers, the required impact strength and ductility are achieved via chemical modification or mechanical blending. PP belongs to the group of commodity thermoplastics and produces a large amount of waste; considerable effort is required to upgrade the characteristics of waste PP in order to recycle it [21]. PP is extensively used in the packaging of consumer goods, which increases the amount of discarded PP packaging material in the waste stream. Postconsumer waste from PP is therefore abundant and

Table 2.2 Physical properties of PP.

Physical parameters	**Value**	**Units**
Specific gravity	0.905	–
Elastic modulus	1,140–1,550	KPa
Impact strength by Izod method at 23°C	65	J/m^2
Hardness (Shore D)	73	–
Melting point (T_m)	178.1	°C
Tensile strength	23.49 ± 0.23	MPa

recycling is a viable option for solid waste management. The recycling of PP is dependent on its structure and properties, and thermal properties are of particular importance when processing waste PP [22].

2.4.1.3 Polystyrene

Polystyrene (PS) is currently widely used due to its good mechanical properties and ease of processing, as well as its relatively low cost [23]. As a result of its chemical stability and ubiquity, a large amount of plastics, including PS, have accumulated in the environment causing a phenomenon termed 'white pollution [24]. New component housing materials made from PS for equipment in the fields of office systems and information technology exhibit a marked improvement in toughness and, with approximately 40% of the market, it is the most important market segment followed by injection molding, electrical and electronics, and refrigerator manufacture. In packaging, the trend is for highly transparent, sparkling and/or tough film.

Table 2.3 illustrates some of the physical properties of PS.

2.4.1.4 Polyvinyl Chloride

Polyvinylchloride (PVC) is the second largest volume of plastic used worldwide and plays an important role in the plastic industry. Due to its compounding versatility, PVC can be processed into many long-life products, such as appliances and pipes. PVC is one of the primary thermoplastic polymers used for the production of biomedical disposable devices owing to its compatibility with a large number of additives

Table 2.3 Physical properties of PS.

Physical parameters	Value	Units
Specific gravity	1.04	–
Limiting oxygen index	17.8	%
Solubility parameter	15.6–21.1	$(MPa)^{1/2}$ in various solvents
T_g	100	°C

T_g: Glass transition temperature.

(i.e., plasticizers, impact modifiers, heat stabilizers), its wide-ranging mechanical properties which yield rigid to flexible end products and, last but not least, the ability to produce relatively low-cost materials [25].

Despite all the environmental debates, PVC is still one of the most important commercially used plastics. Rigid PVC accounts for the major share of PVC production and is used primarily in the construction industry for the manufacture of pipes, fittings and profiles; the outstanding price performance of this material results in the continual displacement of wood and aluminum.

PVC is an integral part of plastic recycling opportunities and can be processed into many short-life products, such as beverage packaging. About 40% of PVC is used for making pipes and recycled PVC material containing virgin grade PVC is an appealing prospect to make new pipe products [25–30]. Table 2.4 illustrates some of the physical properties of PVC. PVC can be thermally cracked with the evolution of hydrogen chloride (HCl) gas which is poisonous and corrosive. Even though it has olefin structure, polyvinylchloride differs from polyethylene only by having every second hydrogen on the chain replaced by an atom of chlorine.

Acrylonitrile-butadiene-styrene

Acrylonitrile-butadiene-styrene (ABS) is one of the most frequently used polymers in the production of electrical and electronic equipment, it also has widespread applications in automobiles, communication instruments and other commodities. Within the electrical and electronic sector, the quantity of recycled plastics could be increased via the recycling of ABS to reduce environmental, economic and

Table 2.4 Physical properties of PVC.

Physical parameters	Value	Units
Specific gravity	1.35–1.4	-
Tensile strength	56.6	MPa
Elongation	85	%
T_g	71	°C
Melting transition temperature	200–300	°C (decomposition occurs)

Table 2.5 Physical properties of ABS.

Physical parameters	Value	Units
Specific gravity	1.01–1.05	–
Tensile strength	6,000	psi
Elongation	5–20	%
Hardness	103	Rockwell R

energy issues. Table 2.5 illustrates some of the physical parameters of ABS.

ABS is sensitive to oxidation due to the presence of polybutadiene components, which act as oxidation sensitizers and lead to the formation of carbonyl groups, which absorb at 1,680–1,750 cm-1 [31].

2.4.2 Polyester

Polyester resins are widely used due to their versatility and economic cost. Polyester exhibits a good combination of resistance to softening and deformation at elevated temperatures, good electrical properties and high resistance to corrosion as well as excellent weatherability [32, 33]. Structural applications, such as reinforced polyester containing glass fiber, comprise more than 80% of the market. Table 2.6 illustrates some of the physical parameters of polyester.

PET has excellent clarity, mechanical properties, and heat resistance. It could be said to straddle the line between commodity and

Table 2.6 Physical properties of polyester.

Physical parameters	Value	Units
Specific gravity	1.26–1.49	–
Tensile strength	6,780–8,860	psi
Elongation	1–13	%
Hardness	86–122	Rockwell R
Melting transition temperature	157–164	°C

engineering plastics. It is used in soft drink bottles, some hot filled food products and in many mechanical applications, and can also be made into fiber and films. Even though polyethylene terephthalate (PET) is low cost, the recycling of wastes of clear or colored PET is limited. High-performance PET is a thermoplastic polyester with excellent thermal and mechanical properties which is widely used as an economical, ecological and consumer-friendly material for video and audio tapes, X-ray films, food packaging, soft-drinks bottles and jars. The demand for PET has increased year-on-year with the subsequent increase of PET waste.

PET recycling technologies have become well known and represent one of the most successful and widespread examples of polymer recycling. There are two methods of recycling PET waste: one method involves physical recycling to produce PET flakes which are reused in a resin, and the other method involves chemical recycling via the depolymerization of PET wastes.

2.4.3 Polycarbonate

Polycarbonate (PC) has excellent mechanical strength, particularly impact strength, good electrical properties and transparency, and is widely utilized in a variety of fields including office machinery, electric and electronic machinery, automobiles, architecture and so on. Many applications require that a PC composition be flame retardant(s) (FR) and combine ease of processing with good optical properties. Table 2.7 illustrates some of the physical properties of PC. Polycarbonate is well known for its balanced properties of rigidity, impact strength, and electrical properties [34].

Table 2.7 Physical properties of PC.

Physical parameters	Value	Units
Specific gravity	1.18–1.22	–
Tensile strength	8,120–9,940	psi
Elongation	4.9–9.4	%
Hardness	70–121	Rockwell R
Melting transition temperature	165–212	°C

Table 2.8 Physical properties of PA.

Physical parameters	Value	Units
Specific gravity	1.03–1.05	–
Tensile modulus	161,000	psi
Flexural modulus	41,800–200,000	psi
Hardness	78–108	Rockwell R
Melting transition temperature	94	°C

2.4.4 Polyamide

Polyamide(s) (PA), the first engineering thermoplastic, was originally developed in the form of high strength textile fibers, which are basically hygroscopic in nature and required drying before processing. Its highly crystalline nature, which can be controlled to some degree during processing, has a major effect on all properties and results in higher strength, higher stiffness, a high heat deflection temperature and so on. PA are sensitive to ultraviolet (UV) radiation and exhibit good resistance to creep and cold flow compared with many less rigid thermoplastics. Table 2.8 illustrates some of the physical properties of PA. Polyamides offer increased long-term durability, flexibility, and resistance to impact loadings.

2.4.5 Biodegradable Plastics

Biodegradation is described as degradation by microorganisms and its chief advantage is that it enables the currently considerable volume of waste to be assimilated by the natural environment [35, 36]. Bioplastics are completely converted into small molecules, such as carbon dioxide, water, minerals and biomass, by microorganisms and do not have any environmental impact or ecotoxicity. The time taken for degradation to occur is measured using the life cycle time. Many polymers that are claimed to be 'biodegradable' are in fact 'bio-erodible, 'hydro-biodegradable', 'photodegradable', undergo slow degradation or just partially biodegradable. The development of a substantial volume of the generated plastics waste, to be commercially viable biodegradable plastics is an important effort towards the preservation and revitalization of our global environment.

Biodegradable plastics are used in tissue engineering and medicine [37]. They are applied as carriers that deliver drugs more specifically to the targe organ and release the drug after a specific stimulus. Biodegradable plastics are an alternative to mostly non-biodegradable plastics to combat the plastics waste problem. However, before promoting plastics recycling, the opportunities and risks of degradable and biodegradable plastics need to be reviewed. Biodegradable plastics has been under investigation as a solution to existing plastics waste problems [38].

2.5 Thermosets

The production of thermoset plastics is quite different to that of thermoplastics; thermoset polymerization occurs via a two-stage curing process using a material and molding technique. It is polymerized via a reaction between two chemicals in the presence of peroxide and an accelerator, with or without heat, and the resulting pressure depends upon the reacting molecules. The majority of linear molecules are formed at the initial stage and even the linear chains contain unreacted portions which are capable of flowing under heat and pressure.

The final stage of thermoset polymerization occurs in the molding press. Under pressure, the partially reacted compound undergoes crosslinking reactions between the molecular chains. Thermoset monomers with three or more reactive ends lead its molecular chains to crosslink in three dimensions. Flexible thermosets have longer chains with fewer crosslinks.

A thermoset plastic has strong, interconnected permanent bonds which are not heat reversible. After curing, reheating does not cause reprocessing, therefore, thermoset materials cannot be remolded. Continuous heating may lead to breaking the crosslinks which in turn leads to degradation.

Many thermosets are polymerized via the condensation reaction in which a by-product, such as water and so on, is created during the reaction in the mold. Volatile by-products cause dimensional instability and low part strength unless they are removed during molding. In some thermoset materials, the reaction occurs via the addition reaction where there is no formation of volatile by-products; permanent crosslinks occur even at RT. Thermosets, due to their crosslinking nature, are resistant to higher temperatures and provide greater dimensional stability than most thermoplastics.

2.5.1 Phenol-formaldehyde

Phenol-formaldehyde resins, the workhorse of the thermosets, are molded via compression, transfer and injection methods, and are principally compounds which contain fillers. They are high-performance engineering plastics used as bonding and impregnating materials, which exhibit good heat resistance and heat deflection temperature, in addition to high electrical and flame resistance properties. It is a low-cost material with excellent foldability and dimensional stability, along with good water and chemical resistance.

2.5.2 Unsaturated Polyester

Unsaturated polyester resins are dissolved in styrene to enable free-radical chain growth co-polymerization between the resin and styrene monomers. Styrene serves as an agent which links the polyester chains resulting in the development of a crosslinking network. Approximately 7% shrinkage by volume occurs during the curing and crosslinking stages [39, 40]. Shrinkage is one of the main disadvantages of this process as it leads to a lack of dimensional stability of the end product as well as poor surface quality. The inclusion of a thermoplastic additive to the formulation decreases the extent of shrinkage. Products containing unsaturated polyester resins are not widely recycled due to poor material characteristics, long processing times, and non-uniform shape and composition.

2.6 Additives

Additives are chemical compounds that are added during polymer production, typically during compounding, to impart desired properties without modifying the molecular architecture of the base material [41]. They are not chemically bonded with the plastics but are dispersed in the matrix. The leaching, omission or chemical transformation of additives results in product failure. Several thousand different additives exist which can be added to plastic polymers in order to create compounds to be used for specific plastic products. Additives, such as antioxidants, nucleating agents and so on, can modify the bulk properties of a polymer and need to be uniformly dispersed in the polymer to be effective. Without exception, all plastics contain additives which are used to impart a variety of functionalities.

Chemical additives are incorporated into plastics to enhance the service lifetime of the end product and improve the processing and physical properties of the material. Different additives modify different properties such as, improving the optical properties, resist aging, enhancing the bulk mechanical properties or for a variety of other reasons. The use of additives is not evenly distributed among the different types of plastic; PVC production alone accounts for 73% of the global consumption of additives by volume followed by 10% in PE and PP, and 5% in styrenics [42].

Additives are commonly used in thermoplastics and thermosetting plastics in order to achieve specific properties, such as stiffness and prevent heat distortion, and to aid processing to produce an economic end product. Plastics contain additives such as heat and light stabilizers, antioxidants, UV-absorbers, lubricants and plasticizers are absolutely necessary for the processing and stability of plastics; in addition, they are required to achieve the desired mechanical properties of the final plastic products.

The selection of additives and compounding process to be used depends on the end applications of the compounded products. Various combinations of additives and stabilizing agents are used during a particular compounding process in order to meet the requirements of a desired application, in addition to preventing thermal oxidation or degradation and discoloration during the melt process, and impart long-term heat and light stability. The presence of additives in the polymer, particularly antioxidants and UV-stabilizers, may reduce the service lifetime of the end product [43, 44]. The migration of additives depends not only on the low

MW and high mobility of the additive but also on the density of the plastic, the additive concentration, contact time and temperature in the system [45], and on the physico-chemical properties of the system components. The number of additives used in the processing of polymers is large and each one has a specific role [46].

2.6.1 Antioxidants

Antioxidants are used as radical scavengers, i.e., to inhibit oxidation by donating a hydrogen atom, in order to compete with the polymer substrate to either form peroxy radicals or breakdown peroxides to form intermediate products, as oxidation reactions regenerate the antioxidant. The use of antioxidant is to combat thermo-oxidative phenomena is a key factor in preserving the polymer's properties. PP is particularly sensitive to oxidation and requires large amounts of antioxidants and UV-stabilizers, as does PE but to a lesser extent [47]. PP may undergo degradation during processing at high temperatures or prolonged use in adverse conditions, such as in the presence of heat, light and chemical agents. Degradation of the polymer results in an increased melt index, lower MW and reduced mechanical properties; therefore, the addition of antioxidants to PP is strongly recommended.

2.6.2 Slip Additives

Slip agents are additives that facilitate the free movement of adjacent plastic sheets in sliding contact, often by causing microscopic surface roughness. Anti-block agents inhibit the sticking (welding) of plastics to each other upon contact or under moderate pressure. The properties that control the effectiveness of amides as slip and anti-block agents have been related to their mode of action [48, 49]. Amides are commonly used as slip agents which help reduce the coefficient of friction and allow the molecules to flow easily.

Amides dispersed in melted PE at RT, at levels exceeding their solubility, migrate to the PE surface during and after the blowing of films. Exuded amides form random aggregates on the film's surface that reduce the close contact and cohesion between adjacent sheets. Slip or anti-block activity increases upon increasing the amount of amide on the PE surface until a critical level is reached; thereafter, further increase of the surface amide concentration does not improve the slip or anti-block properties and may produce an undesirable oily layer on the film. The most effective amides are solids at RT, which may form crystals on the film surface and diffuse rapidly from the PE surface.

The fatty amides used in PE production include: erucamide (EA) (cis-13-docosenoic acid, amide), behenamide (docosenoic acid, amide), oleamide (OA) and stearamide (SA) [50]. Unsaturated amides, such as EA and OA, provide superior slip properties, whereas saturated amides, such as behenamide and SA, provide acceptable slip and superior antiblocking properties. EA displaced OA as the preferred slip agent for the production of polyolefin film because it melts at a higher temperature and is more stable at extrusion temperatures; EA can be used as a slip and antiblock agent [49]. SA also provides useful slip properties, although it is slower acting it has superior antiblock properties to those of EA and OA.

Primary fatty amides are the preferred slip agents and are added at an approximate level of 0.1% during the production of LDPE film materials (in the USA, PE film production was 6.3 billion pounds (lbs) in 1991) [51]. They are less frequently used as antiblock and mold release agents in the manufacture of other flexible and rigid plastic products. EA, OA and SA are very cost-effective for these applications and hence are widely used; however, other saturated and unsaturated amides with a chain length of 18–22 carbon atoms may be equally effective [52].

2.6.3 Ultraviolet Stabilizers

For outdoor applications, additives such as Ultraviolet (UV)-absorbers and light stabilizers are generally used during compounding and are transparent to solar radiation but immune to the destructive effect of UV radiation. UV stabilizers or absorbers are one of the reasons for the long-term non-degradability of the plastics.

2.6.4 Heat Stabilizers

The need exists for thermally stable plastic materials with long-term photolytic stability, which are capable of being compounded into a wide variety of colors. PVC, for instance, requires heat stabilizers to prevent the polymer from degrading during processing, and needs plasticizers to impart flexibility with capable of being compounded into a wide variety of colors [47]. Various thermal stabilizer systems have both advantages and disadvantages associated with them when used outdoors [53, 54].

2.6.5 Plasticizers

Plasticizers are added to polymers in order to increase their flexibility and improve certain processing characteristics [55]. In rigid polymers, adding plasticizer leads to an important increase of processability but to only a marginal increase of flexibility. There are a large number of commercially available plasticizers and include both micro- and macromolecular compounds [46]. The micromolecular compounds are characterized by good compatibility and high effectiveness but have a low permanence due to their migration towards the surface of the product. The omission of plasticizers during plastics processing results in the loss of favorable material characteristics. Using macromolecular plasticizers for the processing of polymers avoids the disadvantages of micro-molecular plasticizers but their effectiveness is much lower [56]. They make material soft and pliable, and examples include epoxidized vegetable oil, butadiene and so on.

Plastics which contain macromolecular plasticizers are difficult to process but result in products of high rigidity, whereas using micro-molecular plasticizers, though a good flexibility and processing is realized, the permanence of the properties is considerably reduced as a consequence of plasticizer migration followed by its evaporation or extraction. As a result, in recent years, both types of plasticizers are usually used simultaneously, with the micromolecular plasticizer content ranging up to approximately 20% [56, 57].

Next to plastics, plasticizers are the most commonly used additive. Major plasticizers, such as dioctyl phthalate (DOP), dioctyl adipate (DOA), dibutyl phthalate (DBP), etc., are added to polymers when flexibility is a major factor in the end use of flexible plastic products such as packaging films, toys, hoses, etc. In plastics film, phthalate esters (PAEs) are used as non-reactive phthalate plasticizers. They are particularly useful in vinyl plastics and are also useful in industrial and consumer products.

2.6.6 Lubricants

The addition of lubricants during plastics processing minimizes the effects of surface heating. Fifteen per cent of the global lubricants and mold release agents market is consumed by the demand for PVC additives. Most developmental work involving lubricants and mold release agents is concentrated on tailoring existing technology for specific applications which is, in part, a result of the relative maturity of the industry. The largest developmental area involves lubricating agents that increase productivity through cleaner-running systems. The goal is to extend the cleaning period beyond one week, to address excess build-up on the tooling. In addition, faster production speeds are being sought by processors and are the product development focus of many suppliers.

As lubricants and mold release agents are frequently used in combination with other additives during the production of PVC, developments occurring in the additives market, such as heat stabilizers and impact modifiers, have a resultant impact on the combination and amount of lubricants and mold release agents used. For example, increasing the use of calcium/zinc heat stabilizers, at the expense of lead and mixed metal stabilizers, has led to the need for better performing lubricants; for lead, polyolefin waxes are generally used, whereas for calcium-zinc, it can be either wax or an ester. Lubricant loading levels have increased and new one packs, which contain all the ingredients for one formulation, are being offered.

2.6.7 Flame Retardants

Twelve per cent of the global flame retardants (FRs) market is consumed by the demand for PVC additives. The PVC production process is a major user of non-halogenated phosphate esters and chlorinated paraffin, in the form of FR/secondary plasticizers, and a minor user of brominated FR. PVC uses significant amounts of aluminum trihydrate or antimony oxide as a synergist which reacts with the halogen contained in the PVC. When FR are added to PVC, they are used as an FR/plasticizer combination; loading levels of the non-halogenated triaryl phosphate esters comprise up to 25% of the total plastic. FR PVC can be used to produce various items such as wire and cable jacketing, items for rigid components to be used in the building and construction industry, flexible films and coated textiles. FR consist of three main components: an acid source, a carbon source and a gas source. FR form a charred-like cellular layer on the surface and this layer acts as a barrier to prevent the transfer of heat to the condensed phase. FR depend not only on their stability but also on their decomposition rate, char-forming rate and char yield.

2.6.8 Mold Release Agents

Mold release agents and processing additives are engineered to improve the production process and product quality. They facilitate optimum demolding of the contents from the mold and the release agent should not transfer from the mold surface to the molded part. Parts that are easily removed from the mold exhibit an attractive mirror-like finish and with less mold maintenance and wear, less cleaning of the part is required.

During the processing stage, every problem that is encountered has many solutions. It is expected that during molding, in particular, the frictional and adhesive forces which contribute to the high level of sticking that arises due to the lack of a mold release agent will constrain material/product formation, resulting in more strain as the required forces developed in the polymer melt would have to overcome these frictional and adhesive forces [58].

Internal mold release agents are mixed directly into the resin to improve the release properties and the molding or processing characteristics. They require heat to function, which originates from an exothermic reaction of the resin, heated mold surfaces or via selected process temperatures. The balance of internal release agent solvent is emitted as vapor, creating the release interface between the part and the mold.

During the resin transfer molding technique, internal mold release helps

- Greatly enhance the resin flow and glass fiber wet out.
- To reduce the use of external release agents

- In contributing to better gloss and improved wet out of reinforcements and fillers
- To reduce porosity in the gel coat.
- To improve the physical properties of the material and enable crosslinking with the resin during curing,
- To improve resin flow with enabled better wet reinforcement.
- To eliminate build-up in the equipment and make mold release easy
- Improve dispersion of colors and fillers with improvement in cosmetic surfaces.
- Eliminate stress and flow marks and yield clear parts with finishing.

In pultrusion, internal release agent helps

- To reduce the pulling force and improve glass fiber saturation during hot and cold press making
- To improve line speeds leading to increased output.
- Eliminate chalking, scaling and scumming

During injection molding, internal mold release agent helps,

- To increase the melt flow index/reduce resin viscosity with quicker production cycles
- Easy mold release with improved dispersion of colors and fillers
- Easy cavity fill with elimination of stress and flow marks
- Improve cosmetic surfaces and keep the parts clean, ready for finishing.
- Reduce the coefficient of friction

External release agents are used to seal the porosity of the mold during the production of fiber-reinforced plastics. A wide variety of mold release agents are known to effectively keep the contents of the mold from adhering to it; however, many of these agents subsequently remain on the surface of the molded product and are extremely difficult to remove [59].

No 'ideal' release agent exists that can provide acceptable non-stick characteristics and also reduce the surface porosity of the molded product. By their very nature, release agents invariably increase the surface porosity, as they reduce the force that attracts the polymer to the mold [60].

Silicon grease and fluorocarbon mould release agent lubricants appear to convert the stick-slip motion into nearly constant slip behaviour, and further reduce the dynamic friction coefficient. This loss of stick-slip motion and reduction in the friction coefficient when using lubricants is considered to be a useful indication for engineers who deal with the process equipment and mechanics [61].

Pentaerythritol tetrastearates are used as external lubricants in PVC and ABS formulations, where they act as mold release agents or anti-blocking agents during the manufacture of plastic films and sheets. Lecithin is also used as a mold release agent [62]. Magnesium soaps are used in the plastics industry as lubricants and mold release agents for thermosetting plastics and thermoplastics. The incorporation of concentrated silicone fluids improve the material flow and act as a mold release agent to overcome the high rate of product rejection. Mold release problems can be overcome by using stainless steel molds with polished surfaces and mold release agents such as silicone oils.

The molding lubricant in the ABS resin should act in two ways to aid processing. First, it should lower the viscosity of the resin and thereby facilitate molding and second, the mold lubricant should serve as a mold release agent as a result of being more concentrated on the plastic surface while solidifying in the mold [63].

2.6.9 Nucleating Agents

Adding a nucleating agent to a plastic is a common way to control the degree of crystallization and tailor mechanical and optical properties [64–66]. The interface between the nucleating agent and the polymer melt has a high surface energy, which reduces the energy barrier associated with the formation of a nucleus. Introducing a large amount of nucleating agent particles therefore greatly increases the nucleation density and consequently the rate of crystallization [67]. In general, a higher nucleation density leads to more desirable material properties; nucleating agents improve hardness, elasticity, optical properties and transparency.

2.6.10 Fillers

Fillers are simple or complex substances of mineral, animal or vegetable origin, which are used for altering plastics properties and lowering cost. These kinds of additives may reduce crack formation and deformation of plastic material, and also improve the surface characteristics, abrasion, temperature and humidity resistance, and flowing properties in cold conditions. At the same time fillers may increase dimensional stability and reduce molding pressure and processing temperature [49, 50, 57]. Fillers improve stiffness, strength and electrical properties; talc, chalk, clay and so on, are used as filling material.

The type and amount of filler have an effect on shrinkage and sink marks; more filler reduces both, but at the same time increases the mixture viscosity, which is critical when working with molding compounds. A combination of coarse and fine particles produces the optimum results. Fillers, such as clay, calcium carbonate and wollastonite, judiciously selected and in relatively high concentrations, can also impart flame retardancy and serve as a stress transfer medium, as well as reduce the total material cost. It should be pointed out that fillers strongly influence the flow characteristics of molding compounds.

2.7 Plastics – Applications

Plastics consumption and utilization has increased in packaging, medical, and all other areas of engineering and consumer products. There has particularly been an increase in plastics consumption in the area of packaging, which has risen to become one of the main pollutants of the environment. Plastics are used in a variety of sectors including:

- Education: computers, mobile communications, fiber optics and the Internet all rely on plastics.

- Healthcare: contribution in medical and biomedical applications helps in many applications such as syringes, artificial muscles, optical lenses, etc.
- Construction: light-weight doors, windows and so on.
- Maximizing natural resources: plastics enable water/food preservation and distribution, using less to do more, e.g., light-weight plastic packaging.
- Energy conservation and efficiency: light-weight plastic vehicle components maximize fuel efficiency and insulation prevents domestic heat loss.
- Renewables: plastics are key to solar and wind power generation.

2.8 Remarks

Plastics are an attractive manufacturing material due to the fact they can be used in a wide range of applications and exhibit high performance characteristics, including good mechanical strength and toughness, strong dimensional stability and processing advantages. A large amount of plastic is used for packaging and container materials as these applications make the most of the unique properties of plastic; the use of any other material for these applications would lead to a three-fold increase in weight resulting in increased transportation and disposal weights, and further environmental loading due to the higher energy demand involved in their production and disposal [68]. Plastics continue to advance into many applications that are used in a range of sectors from automotive and aerospace to medical and electrical. Composites are made from plastic materials by the addition of reinforcing material while compounding.

References

1. Bernardo, C.A., Simões, C.L., Lígia, M., Environmental and economic life cycle analysis of plastic waste management options. A review, Costa Pinto. *AIP Conference Proceedings*, 1779, 140001, 2016.
2. Rebeiz, K., Recycling plastics in the construction industry. *Waste Age*, 23, 35–37, 1992.

3. Hopewell, J., Dvorak, R., Kosior, E., Plastics recycling: Challenges and opportunities. *Philos. Trans. R. Soc. B. Biol. Sci.*, 364, 1526, 2115–2126, 2009.
4. Zheng, Y., Yanful, E.K., Bassi, A.S., A Review of Plastic Waste Biodegradation. *Crit. Rev. Biotechnol.*, 25, 4, 243–50, 2005.
5. Rose, J. and Mayo, F.R., Oxidation of stabilized polypropylenes. *Macromolecules*, 15, 948, 1982.
6. Wims, A.M. and Swarin, S.J., Determination of antioxidants in polypropylene by liquid chromatography. *J. Appl. Polym. Sci.*, 19, 1243, 1975.
7. Sudhakar, M., Doble, M., Murthy, P.S., Venkatesan, R., Marine microbe-mediated biodegradation of low- and high-density polyethylenes. *Int. Biodeterior. Biodegradation*, 61, 203, 2008.
8. Scott, G., Chapter 4, Management of Polymer Wastes, in: *Polymers and the Environment*, p. 80, Royal Society of Chemistry Paperbacks, Cambridge, UK, 1999.
9. Scott, G., Green polymers. *Polym. Degrad. Stab.*, 68, 1, 2000.
10. Epacher, E., Tolvéth, J., Kröhnke, C., Pukánszky, B., Processing stability of high density polyethylene: Effect of adsorbed and dissolved oxygen. *Polymer*, 41, 8401, 2000.
11. Gugumus, F., Physico-chemical aspects of polyethylene processing in an open mixer 5. Kinetics of hydroperoxide formation. *Polym. Degrad. Stab.*, 68, 327, 2000.
12. Gugumus, F., Physico-chemical aspects of polyethylene processing in an open mixer 4. Comparison of PE-LLD and PE-HD with PE-LD. *Polym. Degrad. Stab.*, 68, 219, 2000.
13. Epacher, E., Fekete, E., Gahleitner, M., Pukánszky, B., Chemical reactions during the processing of stabilized PE: 1. Discolouration and stabilizer consumption. *Polym. Degrad. Stab.*, 63, 499, 1999.
14. Scheirs, J., Chapter 5, Advanced Spectroscopic-Based Sorting Methods, in: *Polymer Recycling: Science, Technology and Applications*, John Wiley & Sons, New York, NY, USA, 1998.
15. Knight, W.A. and Sodhi, M., Design for bulk recycling: analysis of materials separation. *Ann. CIRP*, 49, 1, 83, 2000.
16. Chan C.M. (Ed.) *Polymer Surface Modification and Characterization.* Hanser/Gardner Publications, Cincinnati, OH, USA, 1994.
17. Tao, G., Gong, A., Lu, J., Sue, H.-J., Bergbreiter, D.E., Surface Functionalized Polypropylene: Synthesis, Characterization, and Adhesion Properties, *Macromolecules*, 34, 7672, 2001.
18. C.P. Park, Chapter 9: Polyolefin Foam, in: *Handbook of Polymeric Foams and Foam Technology*, vol. 187, D. Klempner and K.C. Frisch (Eds.), Hanser, Munich, 1991.

19. Manias, E., Touny, A., Wu, L., Strawhecker, K., Lu, B., Chung, T.C., Polypropylene/Montmorillonite Nanocomposites. Review of the Synthetic Routes and Materials Properties. *Chem. Mater.*, 13, 3516, 2001.
20. Canevarolo, S.V., Chain scission distribution function for polypropylene degradation during multiple extrusions. *Polym. Degrad. Stab.*, 70, 1, 71, 2000.
21. Karger-Koscis, J., Instrumented impact fracture and related failure behavior in short- and long-glass-fiber-reinforced polypropylene. *Compos. Sci. Technol.*, 48, 273, 1993.
22. Majumdar, J., Cser, F., Jollands, M.C., Shanks, R.A., Thermal properties of polypropylene post-consumer waste (PP PCW). *J. Therm. Anal. Calorim.*, 78, 849, 2004.
23. Singh, B. and Sharma, N., Optimized synthesis and characterization of polystyrene graft copolymers and preliminary assessment of their biodegradability and application in water pollution alleviation technologies. *Polym. Degrad. Stab.*, 92, 5, 876, 2007.
24. Botelho, G., Queiros, A., Machado, A., Frangiosa, P., Ferreira, J., Enhancement of the thermooxidative degradability of polystyrene by chemical modification. *Polym. Degrad. Stab.*, 86, 3, 493, 2004.
25. Dong, D., Tasaka, S., Aikawa, S., Kamiya, S., Inagaki, N., Inoue, Y., Thermal degradation of acrylonitrile–butadiene–styrene terpolymer in bean oil. *Polym. Degrad. Stab.*, 73, 2, 319, 2001.
26. Boo, K.H., Mikofalvy, B.K., Summers, J.W., Alan Sell, W., Mittendorf, D.H., Melt Filtration of Recycled PVC. *Journal of Vinyl Technology*, 14, 3, 140–144, 1992.
27. Summers, J.W., Mikofalvy, B.K., Wooton, G.V., Alan Sell, W., Recycling Vinyl Packaging Materials From the City of Akron Municipal Wastes. *Journal of Vinyl technology*, 12, 3, 154–160, 1990.
28. Dehennau, C., Dupont, S., Benjamin, P., Rijpkema, B., Voituron, G., The Use of PVC Bottle Scraps in Coextruded Light Weight PVC Sewage Pipes, *Proceedings of the Recycle '91*, Davos, Switzerland, April 3–5th April, 1991.
29. Boo, H.-K., Mikofalvy, B.K., G.D., *Emerging Technologies in Plastics Recycling*, Andrews and P.H. Subramanian (Eds.), ACS Symposium Series 513, American Chemical Society, Washington, DC, USA, 1992.
30. Huber, W., Grasl Kraupp, B., Schulte Hermann, R., Hepatocarcinogenic Potential of Di(2-Ethylhexyl)phthalate in Rodents and its Implications on Human Risk. *Crit. Rev. Toxicol.*, 26, 365, 1996.
31. Malik, M., Choudhary, V., Varma, I.K., Current Status of Unsaturated Polyester Resins. *J. Macromol. Sci., Rev. Macromol. Chem. Phys.*, C40, 139, 2000.

32. Odian, G., *Principles of Polymerization*, Wiley, New York, NY, USA, 1991.
33. Hamza, A.A., Fouda, M., El-Tonsy, M.M., El-Sharkawy, F.M., Optothermal Properties of Fibers. I. Relations Between Optical Properties and Change of the Density and Mechanical loss Factor for Annealed Polyester Fibers. Journal of Applied Polymer Science, Vol. 56, 1355-1366, 1995.
34. Albertsson, A.C., Erlandsson, B., Hakkarainen, M., Karlsson, S., Molecular Weight Changes and Polymeric Matrix Changes Correlated with the Formation of Degradation Products in Biodegraded Polyethylene. *J. Environ. Polym. Degrad.*, 6, 187, 1998.
35. Ratajska, M. and Boryniec, S., Biodegradation of some natural polymers in blends with polyolefines. *Polym. Adv. Technol.*, 10, 625, 1999.
36. Atkins, K.E., Chapter 23 - Low-Profile Behavior, in: *Polymers Blends*, D.R. Paul and S. Newman (Eds.), p. 391, Academic Press, New York, NY, USA, 1978.
37. Iman. M., Fathi, M.A., Badr, H., Daly, S., Shirazi, A.N., Dehghani, F., Biomedical Applications of Biodegradable Polyesters. *Polymers*. 8, 20, 2016.
38. Report:Turning today's challenges into opportunities, *A European Strategy for Plastics in a Circular Economy*, European Union, 2018.
39. Atkins K.E., Low-profile additives: Shrinkage control mechanisms and applications, in: *SMC Sheet Molding Compounds - Science and Technology*, Kia HG (ed), Munich; Vienna; New York; Barcelona, Hanser Publishers, 1993.
40. Mascia, L., *The Role of Additives in Plastics*, Arnold, London, UK, 1974.
41. Mailhot, B., Rivaton, A., Gardette, J.L., Photochemical Degradation of Styrenic Polymers, Scheirs, J. and Priddy, D.B. (Eds.), *Modern Styrenic Polymers: Polystyrenes and Styrenic Copolymers*, p. 703, John Wiley & Sons, Chichester, UK, 2003.
42. Meekum, U. and Kenharaj, R., Comparative study of polystyrene foam degradation in the open-air and artificial weathering exposure. *Arabian J. Sci. Eng.*, 27, 1c, 25–32, 2002.
43. Murphy, J., Chapter 2 Types of additives, in: *Additives for Plastics Handbook*, Elsevier Science Ltd, Oxford, UK, 2001.
44. Lacoste, J. and Carlsson, D., Gamma-, Photo-, and Thermally-Initiated Oxidation of linear low Density Polyethylene: A Quantitative Comparison of Oxidation Products. *J. Polym. Sci., Part A: Polym. Chem.*, 30, 493–500, 1992.
45. Ferguson, G.M., Hood, M., Abbott, K., Photodegradable high density polyethylene-based shopping bags—Environmental hazard or blessing? *Polym. Int.* 28, 35, 1992.

46. Gugumus, F., Light Stabilizers, Chapter 2, Zweifel, H., (Ed.), *Plastics Additives Handbook*, 5th Edition, Carl Hanser Verlag, Munich, Germany, 247–403, 2001.
47. Glover, J.H., Slip migration in extrusion coatings. *Tappi J.*, 71, 3, 188, 1988.
48. Luo, N., Janorkar, A.V., Hirt, D.E., Husson, S.M., Schwark, D.W., Coefficient of Friction Reduction of Ethylene-co-Acrylic Acid Film: Effects of Grafted 12-Aminododecanamide and Solvent Exposure. *J. Appl. Polym. Sci.*, 97, 2242–2248, 2005.
49. Molnar, M.N., Erucamide. *J. Am. Oil Chem. Soc.*, 51, 84, 1974.
50. Glenn, G.M., Miller, R.E., Irving, D.W., Chapter 7, Microcellular Starch-Based Foams, In Agricultural Materials as Renewable Resources; ACS Symposium Series; American Chemical Society: Washington, DC, 1996.
51. Swanson, C.L., Burg, D.A., Kleiman, R., Meadow foam monoenoic fatty acid amides as slip and antiblock agents in polyolefin film. *J. Appl. Polym. Sci.*, 49, 1619, 1993.
52. Wirth, H.O. and Andreas, H., The stabilization of PVC against heat and light. *Pure Appl. Chem.*, 49, 627, 1977.
53. Geddes, W.C., Mechanism of PVC Degradation. *Rubber Chem. Technol.*, 40, 177, 1967.
54. Staples, A.C., Peterson, D.R., Parkerton, T.F., Adams, W.J., The environmental fate of phthalate esters: A literature review. *Chemosphere*, 35, 667–749, 1997.
55. Paul, D.R., Chapter 1 - Background and Perspective, in: *Polymer Blends*, D.R. and S. Newman (Eds.), Academic Press, New York, NY, USA, 1978.
56. Krause, S., Polymer Compatibility. *J. Macromol. Sci. Part C Polym. Rev.*, 7, 251, 1972.
57. Van de Beek, A.I.M., Cornelissen, A.A.J., Aalders, T.G., *Fysisch en Chemisch Onderzoek aan Huishoudelijk Afval van 1987, inclusief Batterijen (Physical and Chemical Sudies on Municipal Waste of 1987, including Batteries)*, Dutch National Institute of Public Health and Environmental Protection, Bilthoven, The Netherlands, 1989, Report No.738505007.
58. Soukup, V.G. and Harper, R.C., Molding Rigid Polyurethane Foam. *J. Cell. Plast.*, 8, 144, 1972.
59. Spence, G. and Crawford, R.J., The effect of processing variables on the formation and removal of bubbles in rotationally molded products. *Polym. Eng. Sci.*, 36, 7, 993, 1996.

60. Takeshita, Y., Handa, T., Minami, H., Niwa, S., Sugimoto, K., Dynamic friction property and mechanism of line-type thermoplastic polymer materials for new optical wiring technology. *Polym. Eng. Sci.*, 50, 2258, 2010.
61. Flider, F.J., The manufacture of soybean lecithins. In: *Lecithins*. Szuhaj BF, List G (eds.), Champaign: American Oil Chemists' Society Press; pp. 21–37, 1985.
62. Bair, H.E., The solution of plastics prediction problems by thermoanalytical techniques. *Polym. Eng. Sci.*, 149, 3, 202, 1974.
63. Kristiansen, M., Werner, M., Tervoort, T., Smith, P., Blomenhofer, M., Schmidt, H.W., The Binary System Isotactic Polypropylene/Bis(3,4-dimethylbenzylidene)sorbitol: Phase Behavior, Nucleation, and Optical Properties. *Macromolecules*, 36, 5150, 2003.
64. Kurja, J., Mehl, N.A., Nucleating agents for semicrystalline polymers, in: *Plastics Additives Handbook*, Zweifel, H., (ed.), 5th ed. p. 949 Hanser, Munich, 2001.
65. Menyhárd, A., Gahleitner, M., Varga, J., Bernreitner, K., Jääskeläinen, P., Øysæd, H., Pukánszky, B., The influence of nucleus density on optical properties in nucleated isotactic polypropylene. *Eur. Polym. J.*, 45, 3138, 2009.
66. Wunderlich, B., *Macromolecular Physics*, vol. 2, Crystal Nucleation, Growth, Annealing, Academic Press, New York, NY, USA, 1973.
67. Ozdemir, M. and Floros, J.D., Active Food Packaging Technologies. *Crit. Rev. Food Sci. Nutr.*, 44, 185–193, 2004.
68. Report PIE-No. 39492 – "*Changes and Developments* – Plastic Recycling Today" Duales System Deutschland GmbH, Saarlouis, Germany, 1996.

3

Plastics and Environment

In the physical land mass of the world, countries are territories which are placed as locus in space. The world has countries with cultural and social awareness and self-identity. People perceive and experience their world through the geographical lens of their own country and utilize waste materials according to the recycling methods available. Open spaces in the world should not be set aside for landfills.

Plastics are ubiquitous materials with the advantageous characteristics of hydrophobic materials. Plastics are included in a wide variety of applications which are used in everyday life, as they are light-weight, tough and cost-effective. The main problem occurs at the end-of-life of plastics. The end-of-life decides the best plastics waste management. Therefore, waste plastics are of critical importance.

In the domestic market, plastics are used as packaging materials to retard food products from deterioration and preserve food quality for use at a later time. As industrial parts, plastics help to increase the lifespan of machine parts due to corrosion prevention and wear in comparison to metal parts, which by nature are non-lubricating and susceptible to oxidation reaction. Therefore, plastics have decisively contributed to the standard of living of a society [1].

3.1 Plastics and Conventional Materials – Comparison

The relatively rapid growth of the chemical industry is largely due to the increase of plastics production. On a weight basis, the recent production of plastic and ferrous materials is roughly equal. The manufacture of finished plastic components is increasing by the

Muralisrinivasan Natamai Subramanian (ed.) Plastics Waste Management, 2nd Edition (37–52) © 2019 Scrivener Publishing LLC

same degree of magnitude as competitive products made from other materials.

Reassessment of the competitive position of plastics is a result of the recent dramatic increase in crude oil price, i.e., plastics must be competitive on a price basis as well as their overall technical performance [2]. Table 3.1 illustrates the energy required for the production of PE and other common materials.

Table 3.1 shows the energy required for the production of some common materials on a weight and volume basis, and reveals that the energy necessary to produce 1 volume unit of PE is much less than that of steel. The energy saving realized when using plastics instead of steel or other materials has led to traditional materials being substituted by plastics.

Plastics are more environmentally friendly than any conventional packaging materials. The energy needed to manufacture a paper bag is higher than that needed to achieve the same per-use resource expenditure with a plastic bag. More energy required for the production of other conventional materials means more carbon emissions. A bioplastic material needs to be recycled multiple times, while a

Table 3.1 Energy required for the production of PE and other materials.

Materials	By weight		By volume	
	MJ/kg	**KWh/kg**	**MJ/kg**	**KWh/kg**
PE	21.35	5.93	22.86	6.35
High-density PE	27.32	7.59	25.92	7.20
Low-density PE	53.10	14.75	48.35	13.43
Cement	8.57	2.38	26.82	7.45
Glass	17.39	4.83	45.04	12.51
Steel	30.89	8.58	241.88	67.19
Aluminium	169.09	46.97	446.83	124.12
Copper	55.01	15.28	490.10	136.14

PE: Polyethylene.

cotton bag has to be reused multiple times in order to assure that the environmental impact is similar to a polyethylene bag.

3.2 Effects of Plastics Products and Environment

Environmental impacts from plastics product manufacturing operations can be most effectively characterized by consideration of the different operations involved such as plastics raw material and additives, processing, waste disposal, and recycling or reuse or recovery.

3.3 Landsite Effects

The energy required to break chemical bonds usually comes from ultraviolet (UV) light, stress, heat or pollutants, and oxygen [3]. The main cause of polymer degradation is oxygen, hence an atmosphere lacking oxygen would have a detrimental effect on the degradation of plastics as oxidation would not occur [4]. The weak chemical bonds of polymer chains can also be broken due to poor manufacturing techniques or during the use of the material. Major pollution comes from dumping complete acres of exposed trash, open fires, smoke and rats. During incineration in open air, plastics such as PVC cause the liberation of dioxin.

3.4 Chemical Environment

Sulfur dioxides, hydrocarbons, nitrogen oxides, and particulate matter such as sand, dust, dirt and soot are known to be common air pollutants. These atmospheric pollutants may cause degradation in combination with solar radiation and other weather factors. Sulfur dioxide is a pollutant that arises from industrial sources and soot is acidified by the interaction with moisture and other particles in the presence of sulfur dioxide and oxygen.

Acrylic, nylon and polyester are especially susceptible to hydrolysis and metals, such as iron and copper, which catalyze this degradation. Metal impurities which are present as a result of the manufacturing process, or in a composite material, accelerate degradation. Many plastics are alkali and acid sensitive, and acid can cause the hydrolysis of polymer chains. Organic vapors and fumigants can dissolve or swell plastics.

Plastics such PS, PET, etc., are subject to degradation which is caused or accelerated by chemical agents. Many polymers are actively attacked by pollutants such as ozone, sulfur dioxide and formaldehyde. In addition, water can break the bonds along a polymer backbone present in PET and this hydrolysis reaction is catalyzed when acidic pollutants are present [5].

3.5 Marine Environment

The source of plastics in coastal and open oceans come from refuse or lost fishing gear, waterborne municipal and industrial wastes, and terrestrial litter transported by river systems. Vessels at sea discharge solid wastes such as plastic bags and containers from merchant ships. The most widespread plastic item found in the oceans are raw plastic pellets, which are shipped to plastic fabricators. These beads are introduced directly into the oceans during transport or indirectly by drainage systems that have received the pellets after their loss during manufacture in open oceans. Another vessel-related source of plastics is lost or discarded fishing gear. Entanglement in, and ingestion of, plastics have been documented as the cause of death for numerous sea turtles and seabirds [6]. Discarded or lost drift nets appear to be especially hazardous to certain marine species.

In the 1970s, biologists first reported that seabirds were voraciously eating plastic. Around 50 species of seabirds appear to seek out plastic pellets mistaking them for the small, shrimp-like crustaceans they normally they eat [7]; it has also been reported that fish, turtles, whales and so on, consume plastic. Animals that consume plastic may die of starvation as their digestive tracts become blocked or ulcerated. The toxic chemicals present in some plastics exert a range of effects from the thinning of eggshells to death [8].

Plastics in the marine environment have become inescapable and include everything from fishing nets to tampon applicators to household waste. Commercial fisherman alone purposely dump

52 million lbs of plastic packing into the ocean every year and lose an additional 298 million lbs in the form of plastic nets, lines and buoys [9].

Plastics are manufactured to be durable, which is an important aspect of their resistance in the marine environment. Common plastic wastes are reported to have a life expectancy of 450 years in seawater. Plastic debris in the marine environment is unsightly and threatens tourism at many of the world's beaches. Plastic causes the deaths of perhaps 2 million seabirds and 100,000 marine mammals every year according to the Entanglement Network, a coalition of 18 environmental groups in Washington, DC, USA [10].

In addition to raw plastic, plastic packaging also enters the marine environment in enormous quantities. In 1982 it was estimated that merchant ships alone dump some 639,000 plastic containers into the ocean every day [11]. Most plastics are buoyant and eventually wash up onshore where they slowly disintegrate into small pieces.

The long-term welfare of the marine environment demands greater control over the manufacturing and disposal of plastics. In the overall solid waste stream, plastics are highly visible, particularly as litter on beaches and roadsides. The shipping and fishing industries are a major source of plastic litter in marine environments. An estimated 1 million birds and 100,000 marine animals die each year as a result of ingestion of, or being trapped by, plastics in the oceans [12].

Thrown away plastics lead to other environmental impacts related to marine waste. It is estimated that the amount of plastic waste entering the ocean from land was between 4.8 and 12.7 million tons globally in 2010 [13]. Plastic wastes, therefore, form a major component of marine litter, negatively affecting both aquatic and mainland life. Birds and fish eat food with plastic bags stuck on it. They cannot digest plastic, and either choke on it or die with their intestines packed full of plastic.

Of great concern are thrown away sludge and waste chemicals, which are more dangerous that plastic packaging materials, because of their potential detrimental effects on aquatic ecosystems. These are readily accumulated by aquatic organisms and subsequently transferred along food chains. Therefore, they pose a risk to human health through seafood consumption. The nature of throwaway plastics causes major problems in the ocean. Plastic packaging materials are pollutants of great concern because of their potential detrimental effects on aquatic ecosystems. These are readily accumulated by aquatic organisms and subsequently transferred along food chains. Therefore, it poses a risk to human health through seafood consumption.

3.6 Packaging Materials

Plastic packaging has barrier properties which preserve food products to medicine to consumer products. There have been many applications which have incorporated plastics packaging waste using different technologies based on the nature of the polymer [14–15]. Plastic materials are known to be safe for use with food; however, the finished products cannot have plasticizers or stabilizers in them. Polyolefin materials are used increasingly in foodstuffs and beverage packaging. Some countries are starting to restrict or ban the use of this plastic packaging material because it is extraordinarily persistent in the environment after disposal and is capable of resisting physical, chemical and biological degradation for decades and even longer.

3.7 Agricultural Fields

The use of plastics in agriculture results in an increased accumulation of plastics waste in rural areas. Part of this plastic waste may be recycled, particularly greenhouse films, silage films and fertilizer packaging, sacks, pipes and other plastic products. Other portions of agricultural plastic waste are difficult to recycle for technical and/or financial reasons. In some cases, thin mulching films, thin low tunnel and direct cover films are classed as waste of low recyclability; very thin films are usually heavily contaminated by soil and foreign materials [16]. Furthermore, the use of plastics for mulching in agriculture has increased the plastic litter problem in rural communities.

The most common current disposal practices for non-recyclable agricultural plastics waste are burying in the soil (mulching films), burning, or disposing of it in open fields or landfills [17]. All of these practices have serious negative consequences for the environment, and for the health of farmers and consumers, and impact on the quality and market value of agricultural produce. The process of recovering and recycling agricultural films is difficult as approximately 80% of the weight of the recovered waste plastic is made up of foreign materials such as soil, sand and so on [18–21]. The cleaning of such films results in a higher recycling cost [22].

3.8 Waste Accumulation

The major and main types of plastics are polyethylene, polypropylene, polyvinylchloride and polyethylene terephthalate. There has been an increase in the rate at which plastics waste is being produced, with the production of approximately 150 million tonnes of plastics per year, accumulating in the environment at a rate of 25 million metric tons per annum. PVC is the third most important thermoplastic material with widespread applications ranging from packaging to healthcare devices, toys, building materials, electrical wire insulation, clothes and furnishing [23]. Due to the light weight of plastics, it has become a basic material for all fields, including packaging, the automobile industry, electricity and electronic industries, construction, agriculture, medical packing, etc.

3.9 Degradation of Plastics

3.9.1 Process Degradation

Plastics are exposed to thermo-mechanical degradation during processing involves high temperatures and mechanical shear. During thermal processing, chemical reactions occur at elevated temperatures [24–29]. During the extrusion and injection molding process, the molten polymers experience relatively high rates of shear and hence non-uniform heat generation, which may result in polymer degradation and/or the formation of crosslinks. Thus emission from degradation during plastics processing causes deleterious environmental impacts.

Chain scission may be responsible for the degradation. Chemical reactions may occur with high shear forces and high temperature in an oxygen deficient atmosphere. The usage may also induce physical and chemical changes in morphological structure of the plastics. Therefore, all these introduce structural heterogeneities in the material.

The changes occur in material affects the mechanical properties and stability of the material of the recycling process. The oxidized moieties during processing or physico-chemical changes from the environment and other significant structural changes may enhance the sensitivity in recycling. There is a possibility of further degradation and reduce the long-term stability of the material [30]. It is therefore necessary to have knowledge about the degradation during their life cycle and the effects on macroscopic properties.

Usually, peroxy radicals are formed at the beginning of polymer degradation in the presence of oxygen [31]:

$$RH + O2 \longrightarrow R^{\bullet} + OOH^{\bullet}$$
$$R^{-} + O2 \longrightarrow ROO^{\bullet}$$

The oxidative reaction forms new oxidative moieties due to the utilization of a substantial amount of stabilizers, leading to a consequent decrease of long-term stability and deterioration of mechanical properties. For an effective recycling process, it is necessary to know the extent and effect that degradation has on the structure, mechanical properties and long-term stability of recyclates. Plastics face prolonged use before being discarded as waste [32]. Throughout the life cycle of the plastic, oxidative reactions constantly occur and the new groups formed during these reactions may enhance the sensitivity of the recyclates to further thermal-and photodegradation. Both degradation under the manufacturing process and weathering during the product's use change the properties of the items to be recycled.

Thermal degradation reactions can be highly unpredictable due to various phases have been involved. The reaction behaviour is depend on the characteristic structure of the polymer and still has to be realized. This discrepancy is found in polymer, owing to the complext structural effects, and the various possible effects such as chain fission, scission, etc [33–34].

Polymers are, in the main, hydrocarbons or hydrocarbon-like and as a consequence they are combustible. For many consumer applications, agents to control or suppress this combustibility must be incorporated into the polymeric formulation [35]. Organohalogen compounds, particularly brominated aromatics, continue to be among the most widely used [36]. They function by liberating, upon thermal decomposition, hydrogen halide or halogen atoms which interrupt gas phase flame propagating reactions. Other agents act in the solid phase to promote the formation of a char layer at the surface of the polymer, which protects it from the heat of the flame and prevents polymer degradation and the generation of the volatile fragments which fuel the flame [37, 38].

Polyolefin containing tertiary hydrogen atoms are vulnerable to oxidative degradation via the peroxidation chain reaction [39]. This reaction is accelerated by the influence of sunlight and the elevated temperatures used during processing, and results in the material becoming discolored and embrittled. A continual low level of oxidative degradation eventually leads to mechanical breakdown. PP is a semi-crystalline polymer, the oxidation of which is confined to the amorphous phase.

3.9.2 Environmental Degradation

Environmental pollution by plastic debris was first mentioned in the 1970s [40]. Worldwide plastics production is increasing dramatically with the development of polymers due to their durability, light weight, and cost-effectiveness [41]. Plastics are used in numerous applications, with extensive degradation generating microplastics mostly defined as particles [42].

Natural weathering encompasses the effects of most types of degradation phenomena, which lead to embrittlement or catastrophic failure of the mechanical properties of the plastic after reprocessing [43–47]. Degradation is usually governed by the combined effect of the mechanical properties of the polymer and weathering. Photooxidation takes place on the surface, leading to extensive recrystallisation and the production of surface cracks which can propagate through the product [48]. Embrittlement occurs because of the preferential scission of the tie molecules linking the crystallites in the polymer which causes a disproportionate amount of applied stress [49].

UV light contains energy needed to break the carbon–carbon or oxygen–carbon bonds in the polymer chain. Plastics exposed to UV become brittle or develop a chalky surface as they degrade. Stress can also provide the energy needed to break chemical bonds. Stressed plastics will degrade more rapidly due to the chain-breaking process and are subjected to stress cracking. Microcracks which develop on a UV-irradiated surface provide pathways for the rapid ingress of moisture and chemical agents. Moisture condensation can remove soluble products of photooxidation reactions leading to further degradation.

Certain chemicals and solvents in the environment can aggravate the problem, causing some plastics to crack with minimal stress. Cast acrylic, polycarbonate, polystyrene and PE are subject to stress cracking. UV radiation causes polymer crosslinking and is responsible for the rapid loss of color. The durability of plastics has been considered to be a positive attribute, but it is now commonly perceived by society as negative. Although photodegradability has been suggested as a solution for plastic litter, photodegradation has a limited practical application because UV light will not penetrate landfills [50].

3.10 Environmental Burdens

Plastics waste is complex problem through shared roles and responsibilities among the society as a whole and socially responsible manufacturing and processing industries. Waste is a sign of production inefficiency and it is a non-recoverable cost. In such as complex nature, a growing number of industries are incorporating environmental burdens in their business strategies.

The polymer industry must generate as few emissions during polymerization and consume as little energy as possible at each stage of its life cycle; and the material should be used by the plastics processing industry and not directly by consumers. However, the plastics processing industry generates waste and consumes energy to recycle the product itself, after use by the consumer and disposal of the product at the waste stage.

3.11 Industrial Ecosystem

The underlying principle of industrial ecology is that the industrial ecosystem should not only be as efficient as possible from the environmental perspective, but should also be attuned to the tolerances and characteristics, which requires more efficient use of energy and plastic raw materials processing. It calls for knowledge about the dynamics of natural ecosystems at local, regional, continental and global levels, and a detailed understanding of the capacity of ecosystems to absorb plastic wastes and their recovery period when damaged.

3.12 Remarks

To establish preconditions to adopt a pragmatic approach to environmental improvements, the following could be implemented:

- The government could formulate normative ecological guidelines within which industries ought to act.
- Industries could compare the product chain before and after an environmental improvement has been implemented.

Reduce, reuse, recycle (materials) and recover (energy) are the management strategies in the order of decreasing environmental desirability. The most preferred strategy, reduce, includes source reduction activities, such as replacing single-use products in the market [51]. However, this would not conclude until the environment has reached a sustainable level in absolute terms.

References

1. Bernardo, C.A., Simões, C.L., Lígia, M., Costa Pinto., Environmental and economic life cycle analysis of plastic waste management options. A review. *AIP Conference Proceedings*, 1779, 140001, 2016.
2. Kaminsky, W., Menzel, J., Sinn, H., Recycling of plastics. *Conserv. Recycl.*, 1, 91, 1976.

3. Hawkin, W.L., Environmental Deterioration of Polymers', in: *Polymer Stabilization*, W.L. Hawkin (Ed.), Wiley Interscience, New York, NY, USA, 1971.
4. Grattan, D., The oxidative degradation of organic materials and its importance in determination of artifects. *J. Int. Inst. Conserv. Can. Group*, 4, 1, 17, 1978.
5. Ranby, B. and Rabek, J.F., Chapter 17, Environmental Corrosion of Polymers Causes and Main Reactions–*Environmental Corrosion in Effects of Hostile Environment on Coatings and Polymers*, D. Garner and G. Stahl (Eds.), American Chemical Society, Washington DC, USA, ACS Symposium 229, 1983.
6. Waldichuk, M., Plastics pollution at sea. *Marine Pollution Bulletin*, 17, 12, 529–530, 1986.
7. Dahlberg, M.L. and Day, R.H., Observations of man-made objects on the surface of the North Pacific Ocean, in: *Proceedings of the Workshop on the Fate and Impact of Marine Debris*, R.S. Shomura and H.O. Yashida (Eds.), p. 198, Honolulu, Hawaii, 1984.
8. Merrell, T.R. Jr., Fish nets and other plastic litter on Alaska beaches in: *Proceedings of the Workshop on the Fate and Impact of Marine Debris*, R.S. Shomura and H.O. Yashida (Eds.), Honolulu, Hawaii, p. 160, 1984.
9. US National Academy of Sciences in 'Marine Litter', in: *Assessing Potential Ocean Pollutants: A Report of the Study Panel*, p. 405, Academy Press, Washington, DC, USA, 1975.
10. Freeman, K., We're choking the ocean with plastics, *National Fisherman*, p. 4, 1987.
11. Horsman, P.V., The amount of garbage pollution from merchant ships. *Marine Pollution Bulletin*, 13, 167, 1982.
12. Heneman, B., *Report: Persistent Marine Debris in the North Sea, Northwest Atlantic Ocean, Wider Caribbean Area and the West Coast of Baja California*, US Department of Commerce, Washington, DC, USA, 1988.
13. Jambeck, J.R., Geyer, R., Wilcox, C., Siegler, T.R., Perryman, M., Anthony Andrady, Ramani Narayan, Kara Lavender Law, Plastic waste inputs from land into the ocean. *Science*, 347, 6223, 768–771, 2015.
14. Di Maio, L., Acierno, D., Incarnato, L., Fan, G., Giamberardini, S., Recycled polypropylene for food packaging, Preliminary evolution of the influence of preprocessing treatments and processing. *Ital. J. Food Sci.*, 2, 179, 1995.

15. Lox, F., Van Dael, S., Machiels, S., Organoleptic and migrational properties of pp films produced with various amounts of scrap. *Pack Technol. Sci.*, 5, 307, 1992.
16. Briassoulis, D. and Dejean, C., Critical Review of Norms and Standards for Biodegradable Agricultural Plastics Part I. Biodegradation in Soil. *J. Polym. Environ.*, 18, 384, 2010.
17. Briassoulis, D., Hiskakis, M., Scarascia, G., Picuno, P., Delgado, C., Dejean, C., Labeling scheme for agricultural plasticwastes in Europe. *Quality Assurance and Safety of Crops & Foods*, 2, 2, 93, 2010.
18. González, A., Fernández, J.A., Martín, P., Rodríguez, R., López, J., Bañón, S., Franco, J.A., *Behaviour of biodegradable film for mulching in open-air melon cultivation in South-East Spain*, KTBL-Schrift, pp. 71–77, 2003.
19. Briassoulis, D., Liantzas, K., Hiskakis, M., Management of Agricultural Plastic Waste (APW) - A Pilot Test in Greece, *Proceedings of the European Society of Agricultural Engineers*, Hersonissos, Crete, 23–26th June, 2008.
20. Hiskakis, M., Babou, E., Briassoulis, D., Marseglia, A., Godosi, Z., Liantzas, K., Recycling Specs for Agricultural Plastic Waste (APW) – A Pilot Test In Greece and in Italy, *Proceedings of the European Society of Agricultural Engineers*, Hersonissos, Crete, 23–26th June, 2008.
21. Hiskakis, M., Briassoulis, D., Teas, C., Babou, E., Liantzas, K., Using Agricultural Plastic Waste (APW) as Alternative Solid Fuel (ASF) for Energy Recovery in a Cement Industry Kiln – A Pilot Test, *Proceedings of the European Society of Agricultural Engineers*, Hersonissos, Crete, 23–26th June, 2008.
22. Hiskakis, M. and Briassoulis, D., *Design of Agricultural Plastic Waste (APW) Chain for Greece from Generation to Disposal; A Pilot Test for Energy Recovery, World Congress of Agricultural Engineering*, Bonn, Germany, 3–7th September, 2006.
23. Ghani, S.A., Ismail, H., Yusof, A.M.M., Properties of Recycled Poly(vinylchloride)/ Acrylonitrile-Butadiene Rubber Blends: The Effect of Maleic Acid, in: *Progress in Rubber, Plastics, and Recycling Technology*. 21, 2, 85, 2005.
24. Kimura, M., Reporter, R.S., Salee, G., Blends of poly(butylene terephthalate) and a polyarylate before and after transesterification, *J. Polym. Sci. Part B Polym. Phys. Ed.*, 21, 367, 1983.
25. Robeson, L.M., Phase behavior of polyarylate blends. *J. Appl. Polym. Sci.*, 30, 4081, 1985.

26. Kimura, M., Salee, G., Porter, R.S., Blends of Poly(ethylene Terephthalate) and a Polyarylate Before and After Transesterification. *J. Appl. Polym. Sci.*, 29, 1629, 1984.
27. Godard, P., Dekominck, J.M., Devlesaner, V., Devanx, J., Molten bisphenol-A polycarbonate-poly(ethylene terephthalate) blends. I. Identification of the reactions. *J. Polym. Sci. Part A Polym. Chem. Ed.*, 24, 3301, 1986.
28. Henrichs, P.M., Tribone, J., Massa, D.J., Hewitt, J.M., Blend miscibility of bisphenol A polycarbonate and poly(ethylene terephthalate) as studied by solid-state high-resolution carbon-13 NMR spectroscopy, 21, 1282, 1988.
29. Montaudo, G., Puglisi, C., Samperi, F., Chemical reactions occurring in the thermal treatment of polymer blends investigated by direct pyrolysis mass spectrometry: Polycarbonate/polybuthyleneterephthalate. *J. Polym. Sci. Part A Polym. Chem. Ed.*, 31, 13, 1993.
30. Karlsson, S., Recycled Polyolefins. Material Properties and Means for Quality Determination. *Adv. Polym. Sci.*, 169, 201, 2004.
31. Koverzanova, K.V., Usachev, S.V., Shilkina, N.G., Lomakin, S.M., Gumargalieva, K.Z., Zaikov, G.E., Specific Features of Thermal Degradation of Polypropylene in the Presence of Magnesium Hydroxide. *Russian Journal of Applied Chemistry,* 77, 3, 445, 2004.
32. Vilaplana, F. and Karlsson, S., Quality Concepts for the Improved Use of Recycled Polymeric Materials: A Review. *Macromol. Mater. Eng.*, 293, 274, 2008.
33. Kruse, T.M., Woo, O.S., Broadbelt, L.J., Detailed mechanistic modelling of polymer degradation: Application to polystyrene. *Chem. Eng. Sci.*, 56, 971–979, 2001.
34. Beyler, C., Thermal decomposition of polymers, in: *The SFPE Handbook of fire protection engineering (Section 1, Chapter 12)*, P.J. DiNenno (Ed.), NFPA, Quincy, MA, 1988.
35. Lyons J. W., *The Chemistry and Uses of Fire Retardants*, p15–16, Wiley-Interscience, New York, 1970.
36. Pettigrew, A., Flame Retardants (halogenated), Kroschwitz, J.L, Howegrant M (Eds.), *The Kirk-Othmer Encyclopedia of Chemical Technology*, 4th Edition, Volume 10, John Wiley-Interscience, New York, NY, USA, pp. 954–976, 1993.
37. Green, J., Phosphorus-Bromine Flame Retardant Synergy in a Polycarbonate/Polyethylene Terephthalate Blend. *J. Fire Sci.*, 12, 257, 1994.
38. Green, J., A Phosphorus-Bromine Flame Retardant for Engineering Thermoplastics—A Review. *J. Fire Sci.*, 12, 388, 1994.

39. Billingham, N.C., Calvert, P.D., in *Degradation and Stabilization of Polyolefins*, Allen N.S., (Ed.), pp. 1–28, Applied Science Publishers, London U.K, 1983.
40. Carpenter, E.J., Anderson, S.J., Harvey, G.R., Miklas, H.P., Peck, B.B., Polystyrene spherules in coastal waters. *Science*, 178, 4062, 749–750, 1972.
41. do Sul, J.A.I. and Costa, M.F., The present and future of microplastic pollution in the marine environment. *Environ. Pollut.*, 185, 352–364, 2014.
42. Hidalgo-Ruz, V., Gutow, L., Thompson, R.C., Thiel, M., Microplastics in the marine environment: A review of the methods used for identification and quantification. *Environ. Sci. Technol.*, 46, 6, 3060–3075, 2012.
43. Mallick, P.K., Chapter 6, Design, *Fiber-reinforced Composites: Materials, Manufacturing and Design*, Marcel Dekker, Inc., New York, NY, USA, 1993.
44. Fried, J.R., Chapter 6, Polymer degradation and the environment, *Polymer Science and Technology*, 2nd Edition, Pearson Education, Inc., New York, NY, USA, 2003.
45. Scheirs, J., Chapter 12, Mechanical failure mechanisms of polymes, in: *Compositional and failure analysis of polymers: A practical approach*, pp. 304–351, Wiley, New York, NY, USA, 2000.
46. Paul, D.R., Barlow, J.W., Keskkula, H., in *Encyclopedia of Polymer Science and Engineering*, Vol. 12, 2nd ed., H. F. Mark, N. M. Bikales, C. G. Overberger, and G. Menges (eds.), Wiley, New York, p. 399, 1988.
47. Schnabel, W., *Polymer degradation: Principles and potential applications*. Macmillan Publishing Co. Inc, New York, pp 178–215, 1981.
48. Carlsson, D. and Wiles, D.M., Photooxidation of Polypropylene Films. IV. Surface Changes Studied by Attenuated Total Reflection Spectroscopy. *Macromolecules*, 4, 174–179, 1971.
49. Oswald, H.J. and Turi, E., The deterioration of polypropylene by oxidative degradation. *Polym. Eng. Sci.*, 152, 6, 1965.
50. Scott, G., Photo-biodegradable plastics: Their role in the protection of the environment. *Polym. Degrad. Stab.*, 29, 135, 1990.
51. Hayrabolulu, H., Şen, M., Çelik, G., Kavakli, P.A., Synthesis of carboxylated locust bean gum hydrogels by ionizing radiation. *Radiat. Phys. Chem.*, 94, 1, 240–4, 2014.

4
Plastics Processing Technology

Production processes have become more efficient due to globalization, and it is important for plastics used in world markets to be sufficiently flexible for production of a variety of products. Plastics processing allows integration to be accomplished effectively and efficiently. Plastics are not ideal solids; they exhibit, in common with all other materials, elastic properties that are combined with viscous or flow properties. Hence, they are called viscoelastic materials.

The term "plastics processing" indicates that plastic raw materials conversion systems into marketable products. The business response to waste recycling with plastics processing has led to an emerging interest towards materials management. This is considered to be a fundamental aspect which creates a huge challenge for plastics waste management.

4.1 Background

Plastics processing emerged as a separate, identifiable engineering discipline in the late 1950s and 1960s. The most exciting developments can be attributed to important advances which took place in the late 1980s and 1990s, which include [1]:

- Catalysts were used in the polymerization process to improve the material characteristics, particularly in the case of polyolefins.
- Progress in the field of high-performance plastics.
- Production of polymers with improved properties as a result of scientific compounding.

Muralisrinivasan Natamai Subramanian (ed.) Plastics Waste Management, 2nd Edition (53–72)

- Research and practices were enhanced by enormous computational power due to the computer revolution.
- Instruments to measure quantitative properties during flow and deformation, and physical properties to the molecular level as well as determination of the chemical composition of the material.

4.2 Management – Plastics Processing

Plastics are of importance in all areas of our daily life. They are fabricated in very large ranges of system, volume, weight, and construction. In between a huge variety of plastics exists products designed for all kinds of applications by plastics processing technology. To meet all the needs of the market, plastics offer a large range of optimized barrier, mechanical, and chemical properties, all at an economic price. Plastics processing techniques are an economically viable and versatile way to produce new materials with a wide range of properties by merely using processing equipment such as injection molding, extruders, or internal mixers.

Plastics processing started with the development of thermoplastics. Traditional processing techniques, such as extrusion, injection molding, thermoforming and rotational molding, have, and continue to, benefit from significant technological progress [2] not only with virgin materials but also recycled materials. In plastics processing management, the optimum conditions depend on machine size, resident time of the material in the machine and other features of the process. The waste plastics are explored through an experimental approach in processing.

Processing often requires external heat for the melting step and results in frictional heat which impacts the quality of the product. During processing, plastic materials are usually in a molten state and exhibit high viscosity. The melt flows through complex geometries and undergoes viscous dissipation, shrinkage and warpage to take the shape of the mold after deformation. The level of viscosity adds more challenges to the processing and shaping of plastic materials during the stages of deformation, flow and solidification of materials. Viscosity plays a central role during processing until the final stage when the material cools down and solidifies into the final product. During the molding of plastics, temperature is one of the most important parameters and temperature distribution has a significant effect on the properties of the final part. An important aspect of plastics processing management is to obtain products with the required end-use properties such as melt viscosity, softening temperature, mechanical strength and so on.

Therefore, plastics processing depends on:

- Material characteristics
- Temperature settings
- Specific output per screw revolution
- Melt temperature
- Melt pressure
- Machine output

The most significant costs for the plastics processing industry are input and labor costs and the value of output. In general, industries have been unable to increase the price of products to match the rise in costs—labor costs, in particular—and have considerably increased recycled plastics and fillers such as calcium carbonate to reduce the prices. However, sometimes the recycled plastics create trouble during processing with an increase in ultimate waste with virgin material. Additional problems have been the fluctuations in the price of oil, which is an important determinant of the price of plastics.

4.3 Plastic Materials - Variations

Plastics are concerned primarily with the liquid (viscous) and solid phases during their processing and molding respectively. In crystalline materials, the change from solid to liquid is abrupt and easily discernible. In an amorphous plastic material, the change is not abrupt or readily apparent, such as water to ice or ice to waste.

- The material used is not identical each time, having different heat histories, molecular weights, molecular

weight distributions, degree of polymerization, and impurities.
- The material is not linear in most of its functions.
- It may differ according to the direction of the material flow in the mold.
- It is exposed to moist air and compressed with it in the heating cylinder, with varying amounts of thermo-oxidation with changes in pressure.
- The material is not linear in most of its functions.
- It is compressible, stretchable, elastic, and subject to changing properties after removal from the mold.
- It has time-dependent properties that are strongly altered by its environment.
- In some materials, there are varying degrees of crystallinity, which are not reproducible or predictable.
- There is no dramatic visible change in its properties at any given point.

For these reasons, quantitative solution in reprocessing are not yet possible.

4.4 Technology

Plastics processing can shape material and improve the properties of end products. The use of plastic enables parts to be manufactured which have various characteristics including: light-weight, high strength, high precision, high efficiency and low cost. The development of packaging and other consumer products, in addition to automobile, aerospace and hi-tech industry parts, all include plastics processing.

Plastics processing can be distinguished as production related to different segments, namely

- Plastic plates, film, tubes, and shaped sections. Plates and films that are subsequently processed further within the branch.
- Packaging materials such as crates, bottles, etc.

- Construction materials such as window frames, wall coverings, etc.
- Medical and health care products such as syringes, blood bags, etc.
- Other parts such as technical parts, household articles, automobile parts, electrical parts and electronics, etc.

The processing of thermoplastics or recycled plastics mostly involves changing pellets, granules and powders into different forms. The type of polymer determines the structure–property relationship of various physical properties, i.e., the properties of the material dictate the complex structure of plastics [3]. The commonly used techniques for product fabrication are:

- Injection molding
- Blow molding
- Extrusion
- Thermoforming
- Rotational molding
- Compression molding

The processing technique to be used depends on the material type, product shape and size, production requirement, accuracy and quality, and design load performance. Injection molding is used to form different shapes and sizes of intricate design. The extrusion process is used to manufacture pipes, sheets, profiles and so on, during a continuous process. Blow molding and rotational molding are used to manufacture hollow parts. Thermoforming transforms extruded products in sheet form into various shapes and sizes. Compression molding is used to manufacture thermoset products and mixed plastics waste.

Processing techniques involve:

- Mixing, melting and plasticizing
- Melt transporting and shaping
- Drawing and/or blowing
- Finishing

The choice of processing method employed when managing plastics waste material depends on the time and money available. As there are numerous ways to consider shaping a plastic melt, the cost and capital involved tend to dictate the available options of plastic processing methods. Plastics processing involves using molds or dies and processes such as injection, extrusion, calendering and so on, which are all aspects of processing that have to be managed.

The most suitable plastics processing technique used for the production of any component will depend upon the shape, size and quantity required; in addition, the cost of plastics waste, obtained by different processing methods, must be taken into account. The product may need to be altered to make it suitable for an individual process and may even include material change. Many processes are available for the production of plastic products and the choice depends upon the relationship between material properties, processing method and end-product properties, in addition to the choice of forming method, which all have technical and economic aspects. Three important features of plastic melts during processing are shear viscosity, melt flow index (MFI) and melt elasticity. MFI is the quantity of polymer extruded under specific load and temperature conditions in a given time. The elastic properties of the melt are a major factor in determining the residual strain and molding defects [4].

4.4.1 Injection Molding

Injection molding is one of the most widely employed methods for the manufacture of plastic products. Injection molding can be achieved using either a reciprocating-screw-type machine or a plunger-type machine. Reciprocating screw injection molding has been commercially exploited on a large scale.

During the manufacture of plastic products there are five basic stages involved in injection molding: plasticization, injection, packing, cooling and finally ejection. During the plasticisation stage, the raw material in solid form is transformed into molten material through the combined action of the friction provided by a rotating screw and the heat provided by external heating elements. The retraction of the screws during rotation forces the molten material to accumulate at the front of the screw. When a sufficient amount of material has accumulated, the screw moves forward, causing the molten material to fill the cavity. During filling, the hot polymer melt rapidly fills a cold mold, reproducing a cavity of the desired product shape. Figure 4.1 illustrates the stages of the injection molding process.

Once filling is complete, additional material is packed into the cavity in order to compensate for the shrinkage which occurs during the cooling stage. During the packing and holding stage, the pressure is raised and extra material is forced into the mold to compensate for the effects of temperature decrease and crystallinity development, which determine the material density during solidification. The cooling stage starts with the solidification of a thin section at the cavity entrance or exit of the mold impression and holding pressure is released. When the solid layer on the mold surface reaches a thickness sufficient to assure the required rigidity, the product is ejected from the mold [5].

After solidification, the molded article is ejected and the sequence of events is repeated in a cyclic manner. The cooling time constitutes a major portion of the molding cycle due to the poor thermal conductivity of plastic materials. The choice of mold and material delivery system will depend on the characteristics of the part being molded as well as on the material. Among plastics processing techniques, injection molding is an attractive option in many emerging plastics such as polyethylene, polypropylene, etc. Their use can help

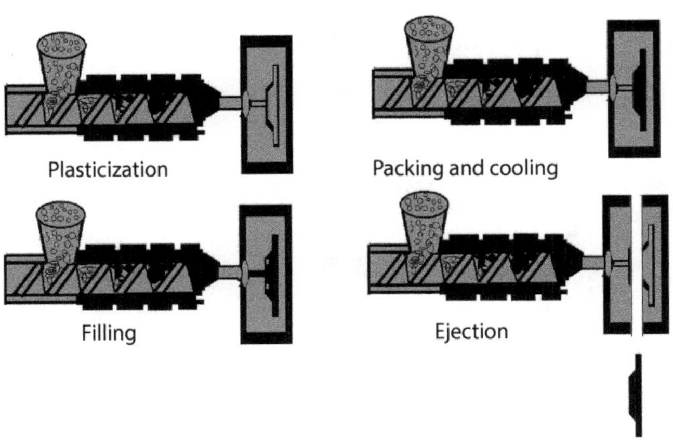

Figure 4.1 Stages of the injection molding process.

solve the problem of increasing commodity waste plastics. Thus, the rheological properties, cycle efficiency and capacity can be maintained with the help of processing temperature and pressure.

4.4.2 Blow Molding

During extrusion blow molding, the material in the form of a pellet or granule, is fed into a plasticizing extruder. The plastic is melted via external heating and the hot melt is transferred through the barrel by the shearing motion of the extruder screw, which has solid conveying, melting and metering sections to ensure a uniform homogeneous melt at the screw tip. The helical flights of the screw configuration change along its length from input to output.

The screw feeds the melt directly into the head-die assembly within a continuous extrusion blow molding machine. The melt flows around the mandrel and into an annular die of either convergent or divergent type. Parison, which is a hollow tube, can be continuously extruded and the extruded hollow tube is cut at preset time intervals for transfer into the blow molding stage. Figure 4.2 illustrates the stages of a continuous extrusion blow molding machine.

In a ram- or plunger-type machine, the extruder feeds the material to an accumulator/head device. The plunger pushes the material rapidly through the head-die assembly once the desired volume has accumulated. The transfer of material does not need to occur via a blowing station in the mold clamp mechanism. The part is blown, cooled and removed from the mold, and the next parison is only extruded after the part has been removed.

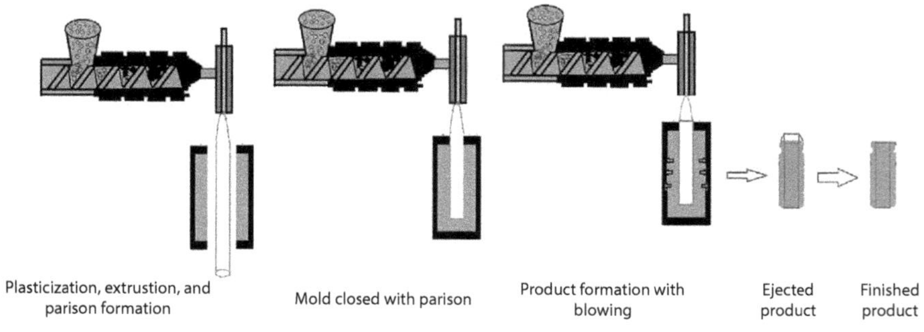

Figure 4.2 Different stages of continuous extrusion blow molding.

When the desired length of the parison is reached, the mold is closed and the parison is inflated by internal air introduced via the die-head assembly. The mold walls are vented and a vacuum may be applied. The molten plastic is thus forced to conform to the shape of the mold cavity. The article is then cooled, solidified and ejected from the mold.

The annular die can be designed to incorporate a hydraulic mechanism to vary or program the annular gap size in both extrusion- and plunger-type blow molding. With extrusion blow molding, a specific wall thickness distribution or controlled weight of the parison can be programmed.

Injection-stretch blow molding is a two-stage process. The material is injection molded around a core rod to create a preform during the first stage. The preform is then stretched through the action of the core rod, inflated and cooled, similarly to the extrusion blow molding process. Injection-stretch molding yields lighter products with biaxial orientation and increased tensile strength.

Figure 4.3 illustrates the various stages of injection-stretch blow molding. The biaxial orientation also provides permeation to gas, liquid and odor due to an increased molecular alignment. This process produces bottles or containers which are scrap-free, and have close tolerance, i.e., a complete finish which does not required any secondary operations.

It is can be possible to recycle in-house or select source of plastics waste (blow grade). This technique is not very useful with either mix plastics or with contaminants.

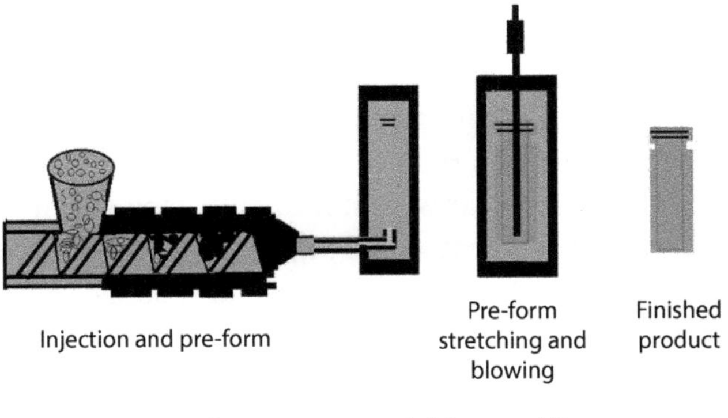

Figure 4.3 Various stages of injection-stretch blow molding.

4.4.3 Extrusion

The extrusion process is usually divided into three stages: solid conveying, melting or plasticization, and melt conveying or metering. An extruder consists of a threaded shaft or screw rotating in a fixed cylinder or barrel, plus a feed port and a die. The screw channel constantly changes its position relative to the barrel and the advance of the polymer melt through the barrel of the extruder follows a helical path, which is a mirror image of the extruder screw. During the extrusion process, pellets or granules are compacted and fluxed in the extruder barrel with the help of external and friction heat. The molten material is extruded inline under pressure through a die opening designed to yield the required pipe, profile or sheet. The hot extrudate is immediately passed through a calibrator which has a water circulation channel in a vacuum environment. Passing through the water channel, the extrudate develops sufficient strength to be pulled away from the die by a suitable puller or take-up equipment. The resultant extruded product is either cut to length or reeled as required.

Dies should be heated using electrical band heaters which are specially shaped to give complete and close conformation to the outside die dimension. The die opening should be 20–30% oversized to accommodate the drawdown caused by the constant tension necessary to draw the molten material away from the hot die and is commonly known as the swelling ratio. The extruded product is normally cooled by immersion in water, water spray or a combination of the two.

Figure 4.4 illustrates the different types of extrusion processes. In the blown film process, extrusion is carried out using standard side- and bottom-fed die types, both rotating and stationary. Rotating dies are preferable because of their ability to minimize gauge bands. Conventional bubble-cooling methods and take-off equipment are used to cool and form the film.

Due to process/product requirements, extrusion requires plastic material of higher molecular weight for better melt strength. During extrusion, the material is heated and pumped into a permanent mold, which is called a die, where it takes shape and cools. The extrusion process involves continuous material heating, shaping and cooling. The screw design plays an important role and optimum processing conditions maximize product quality.

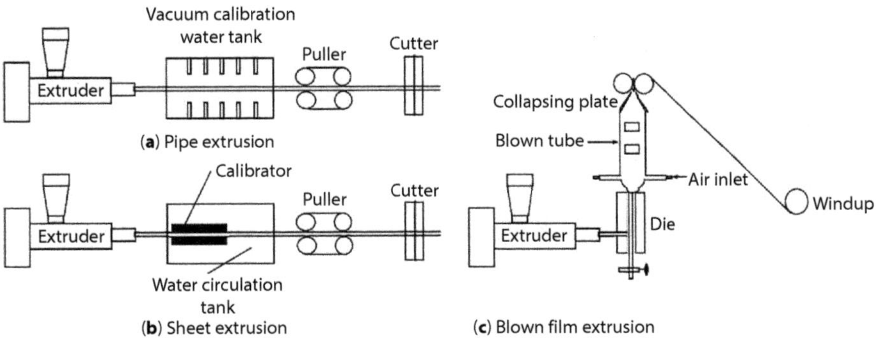

Figure 4.4 Different types of extrusion (a) pipe extrusion, (b) sheet extrusion and (c) blown film extrusion.

4.4.4 Thermoforming

Thermoforming is a process of deformation by heating the plastic sheet to its rubbery solid state. A plastic sheet is heated to a suitable temperature in order to undergo deformation by temporarily draping the softened sheet over a rigid mold until the material takes on the desired finished shape; this process includes a variety of ways to produce rigid articles from a flat sheet. Of special concern during the thermoforming process is the thinning of barrier materials at high draw ratios, with the potential loss of barrier properties.

Plastic materials have low thermal conductivities, hence during heating, even relatively thin sheets can have surface temperatures that are substantially higher than the interior temperatures. Therefore, plastic sheets must be heated to an optimum forming temperature in order to achieve uniform and repeatable drawdown. Figure 4.5 illustrates continuous sheet forming extrusion via the thermoforming process.

Common stages of heating thin-gauge sheets involve conduction, convection and radiation. During the conduction stage, the energy

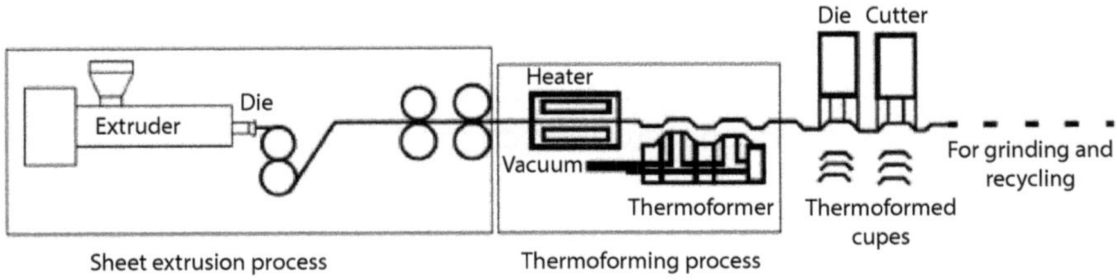

Figure 4.5 Continuous sheet forming extrusion *via* the thermoforming process.

interchanges between the layers in multilayer sheets. During convection, heat transfer occurs between the hot air and the sheet. During thermoforming, the radiation energy is between the electrical oven elements and the sheet, and the preferred choice is thin-sheet thermoforming.

4.4.5 Rotational Molding

Rotational molding is one of the plastic processing methods for the production of hollow plastic products. Material of recycled plastic powder is charged and subjected to biaxial rotation in an oven at a temperature of 200–240°C. The plastic powder melts inside the mold by heat transferred *via* convection through the mold wall. All the powder melts and the mold is moved out of the oven while maintaining the biaxial orientation. Exposure to air, or using a fan or water shower are usually employed to cool the mold.

Once the product has reached sufficient rigidity in the mold after cooling, preferably below the melting temperature of the plastics, the mold is opened and the product is removed. The rotational molding process involves material charging, heating and rotation, cooling and rotation, and finally demolding. Figure 4.6 illustrates the rotational molding process.

The rotational molding process requires a long cycle time compared with other process such as injection and blow molding. The main attraction of this process is that in addition to being an alternative to injection and blow molding processes, it is inexpensive and can handle complex shapes with variable thickness and a wide range of part sizes. By rotating the mold, heating and cooling are commonly achieved by convection to the external surfaces of the mold using air as the transfer medium.

Rotational molding depends on plastic powder transport, heat and mass transfer within the equipment. Temperature is the cornerstone of the molding process. The residue or ash residence time predicts the degradation or decomposition that occurs in the incinerator. With mix recycle materials, the main factors affecting the products are

- Complex nature of plastics composition
- Physical properties
- Rheological properties

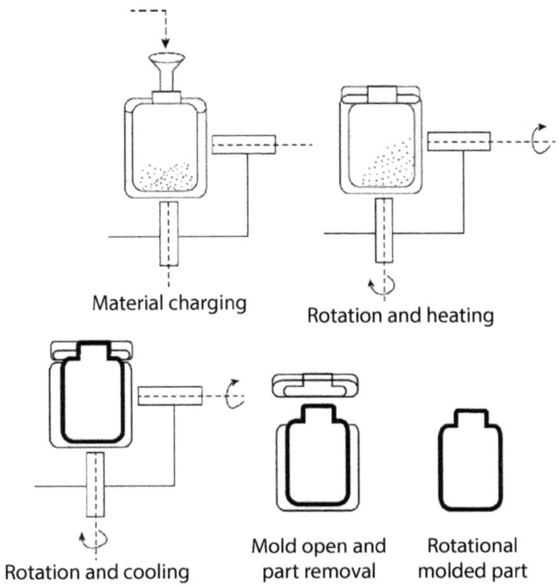

Figure 4.6 Rotational molding process.

The main factors affecting rotational molding of waste plastics are the axial direction where the density separation occurs, the weight fractions present in the composition, rotation speed, phase coherence, and the angle of response.

The axial motion of powder plastics waste inside the mold depends on the mass flow rate of solid particles, weight of the material, rotational speed, temperature and inclination of mold. Axial velocity of the powdered material varies along the mold length and causes thickness variations in the molded articles. Most of the molds developed are based on the assumption of constant height throughout the length of the rotational mold. Additives influence rheological properties present as solid materials in plastics.

The concept of rotational molding (Figure 4.7) with respect to waste plastics, the heating time and temperature are interrelated. The increase in temperature beyond the processing of temperature or increased time will degrade further the material. Not only degradation, there may fire hazard. To prevent fire during waste plastics processing, it is essential to have the flame retardant additive to be added and thereby waste plastics powder is to be managed in production and waste utilization.

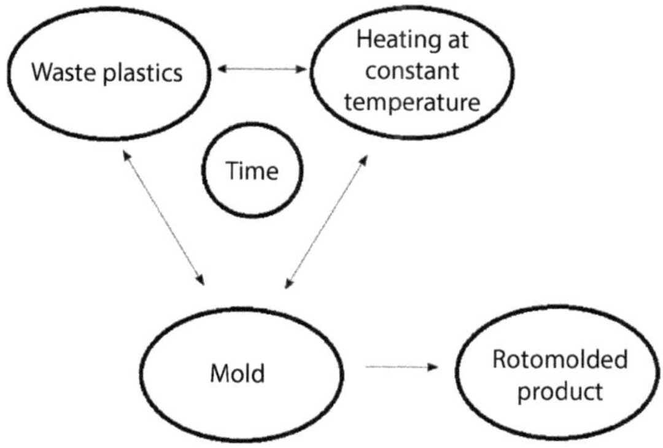

Figure 4.7 Concepts of rotational molding.

4.4.6 Compression Molding

Plastics waste is hydrophobic in nature. It can be possible to blend it into composites, which lowers the environmental impact. It is difficult to achieve a recycle level that plays in the global waste plastics market. Manufacturing of plastics composite product from mix waste or contaminated plastics or recycled plastics is very simple with known technology, machinery and equipment used in plastics processing industries. Therefore, implementation of manufacturing the plastics composite into products does not require much investment.

Mix waste or contaminated plastics are common in municipal waste plastics. These materials can be compressed using compression molding technique in order to improve mechanical and physical properties of the material as composite. It is used to reduce the cost of sorting of waste plastics into separate materials.

Blending mixed or contaminated waste plastics is a very cost-effective and efficient method for recycling plastics waste. However, it is important for plastics to adhere one to one to form co-continuous type morphology to provide the necessary strength properties. Waste plastics are multiphase systems which are known to have poor interfaces and exhibit some incompatibility for at least 60–90 percent of the available waste. Therefore, it is necessary to modify their interfaces through the use of compatibilizers.

Compatibilization of mixed or contaminated waste plastics underlies the basic principles of mechanical, non-reactive or reactive types. This compatibilization of plastics waste is to be made as value-added products, such as panels, doors, etc., using the compression molding technique. The unknown composition of compression molded sample indicates that the material's strength is significantly less than that of standard plastics or their copolymers and blends [6]. The compression molding technique is used to recycle mixed or contaminated waste as it is a cost-effective technique in comparison with other methods to reduce pollution such as mixing with bitumen, construction, etc. [7].

Compression molding is an attractive and effective method of modifying the properties of polymers. Mixed waste plastics with different physical properties are achieved through the synergistic effect of the desirable properties of each component in the mix. However, most mixed waste is immiscible due to the unfavorable enthalpy of the polymers present in it. The phase separation and weak interfacial adhesion as a result, greatly limit properties. The morphology plays an important role in determining the ultimate mechanical properties. In mixed plastics prepared by melt processing, morphology development is controlled by the interfacial tension between the plastic materials. Compatibilizers tailor the interfacial tension and refine the phase morphology. Interfacing leads to lower interface tension and benefits the stress transfer across the interface [8–10].

4.5 Productivity and Task

Productivity improvement is the primary objective of the plastics processing industry and developments include the entire process. The results of these developments have been implemented in terms of the utilization of plastics waste, additives and machinery. The role of plastics processing management has expanded, from simply executing to implementing, and requires industries to continue to provide and develop efficient systems and techniques. Improvements are driven by respective solutions to process problems and the development of productivity can be achieved by implementing further waste reduction, which is the primary aim of the processing industry.

- Injection molding, extrusion, calendaring, etc., are fast, inexpensive and easily controllable; however, recycling of in-house waste plastic material will be suitable for

processing, until otherwise, small contaminant or mix of other material will upset the whole process and generate waste material which further contaminates the process.
- Rotational molding is suitable for small contaminant of the polyolefin family. The waste mix between polyolefin is not greater than 2%. In the case of two or three layers, the inner layer can be used with contamination of other material not more than 5%.

The development of a variety of materials and products based on the blends from waste plastics mixing can be achieved with proper proportion of plastics materials to improve the mechanical strength [11]. These blends can be used in extrusion and other suitable techniques. This is reflected in the development of a variety of materials and products based on plastics blend from waste plastic materials.

4.6 Waste Processing

Plastics waste processing, in particular, has direct links with the conservation of energy and natural resources, and the development of technology is the best solution to deal with plastics waste material. Plastics processing is used to recycle waste, which is driven by the potential depletion of raw materials and socio-economic factors. However, the direct reprocessing of plastics waste is limited due to contamination or the mixture of different kinds of materials.

The waste plastics used may not be identical each time. They may have different heat histories, molecular weights and distributions, degrees of polymerization, and impurities. Therefore, in reprocessing or recycling, the processing variables, such as the processing machine, change conditions in order to compensate manually for changing conditions. The formation of granules or pellets offers

- Plastics waste during melt processing that produces more homogeneous materials.
- Shear during melt mixing that makes material flow more easily, with most of the heat being supplied directly to the material to melt.

Reprocessing is important and is usually determined by trial and error. The factors that affect reprocessing or recycling are:

- Material temperature
- Temperature profile
- Pressure
- Material velocities
- Viscosity
- Flow patterns

Material waste, such as nylon, polycarbonate, and cellulose acetate, are hydroscopic. They must be dried before reprocessing. The moisture causes surface defects and seriously degrades the material. The drying is to be carried out under controlled and circulated heated air.

Plastics waste processing depends upon technology. The overall share of processing waste in combination with virgin material is negligible in plastics processing management. Processing waste has a strong focus on experimental research as the use of 100% recycled material may lead to a deterioration in material properties. Current development activities are focused on the processes and practices of reprocessing plastic waste materials, with emphasis on the processes and practices that reduce emissions and integrate the life cycle assessment (LCA) of technologies and materials. Plastics processing management involves the reuse of plastics waste from industry, and primary and secondary recycled material, to limit the environmental impact and preserve natural resources.

4.7 Reprocess Material in Plastics Processing

Mixing reprocessed or recycled material with virgin material has an effect on structural and macroscopic properties. A small addition of process waste material with virgin material optimizes the consumption of industrial waste generation during processing. Not only are virgin plastics exposed to thermo-mechanical degradation, but also waste plastics during processing. Structural changes may enhance the sensitivity in recycling. Chain scission may be responsible for the degradation. Chain scission induces the molecular weight reduction, thereby increasing the degree of crystallinity in semicrystalline polymers. It decreases the viscosity and change in mechanical properties, which results in progressive embrittlement of the recycled material. Therefore, structural heterogeneities are introduced in the material. The changes that occur in waste material affects the mechanical properties and stability of the material of the recycling process. There is a possibility of further degradation and reduction of the long-term stability of the material [12]. It is therefore necessary to have knowledge about the degradation during its life cycle and the effects on macroscopic properties.

The machinery is designed to circulate the waste generated during processing in normal processing for quantity addition with virgin material.

- In biaxially oriented plant (BOPP), the trims which generate during processing are circulated into the continuous processing. The trims are passed to film grinder. The crushed film waste is pumped into the crammer feeder where the waste and virgin material mix in proportion.
- In PVC processing, the chain scission, crosslinking, and chain branching occur while recycling without addition of stabilizers and other additives. Therefore, processing condition and addition of stabilizer and additives is essential to optimize the processing of recycled PVC.

Plastics are assessed with multiple recycling by processing homogeneous waste fractions. Some of the sources of homogeneous waste are agricultural film waste and industrial scrap. Optimization of processing conditions during mechanical recycling is important to avoid further waste. Resource efficiency is related to plastics manufacturing from raw material to end products using raw materials, recycling to manage production and waste utilization. Waste plastics processing also includes products performance not to become waste, environmental performance, recyclability, durability, and environmental compatibility of plastics material.

4.8 Challenges and Opportunities

As the production and processing of thermoplastic materials occur with the material being in a molten state, knowledge of the flow behavior of the material is essential [13]. A complete understanding of the plastic process from production to the properties of the end product has become increasingly important in view of the increasing speed of material development resulting from rapidly changing customer requirements [14]. To achieve the desired end-product characteristics, plastic processing technology depends upon a balance of material selection, processing parameters and mechanical property requirements; in addition, an important factor of plastic production is its ease of processing. Plastics processing faces challenges and opportunities in order to develop advanced techniques, and innovative materials are needed to achieve this; hence, plastics processing requires hi-tech involvement and extending current knowledge [15].

4.9 Remarks

Plastics processing industries recycle most of its industrial waste internally. Recycled or multi-recycled products is to have a lesser strength to fit in suitable application not with much mechanical properties. Compounded waste plastics ready for fabrication, molding, rolling or blending. Many of the scrap plastics are from repackaging, off-grade, fabrication scrap, contaminated plastics and mixed with colors. Plastics separation with wet and dry methods provides separate reuse plastics. Technologies require for waste segregation, grinding and air separation to process plastic materials.

The output of the plastic processing industry increased compared to overall manufacturing output. Output growth is expected to slow down due to single-use plastic materials, although once again it is likely expected to exceed that of the manufacturing industry as a whole, and this is particularly true of technical parts, packaging and semi-finished products.

References

1. Tadmor, Z., The evolution of polymer processing into macromolecular engineering and science. *Plastics, Rubbers and Composites*, 33, 1, 3, 2004.

2. Vergnes, B., Vincent, M., Demay, Y., Coupez, T., Billon, N., Agassant, J.-F., Present. Challenges in the Numerical Modeling of Polymer-forming Processes. *Can. J. Chem. Eng.*, 80, 1143, 2002.
3. Chum, P.S. and Swogger, K.W., Olefin polymer technologies—History and recent progress at The Dow Chemical Company. *Prog. Polym. Sci.*, 33, 797, 2008.
4. Technical Report, Plastic Processing, Materials & Design, 3, 5, 566–570, 1982.
5. Pantani, R., Coccorullo, L., Speranza, V., Titomanlio, G., Modeling of morphology evolution in the injection molding process of thermoplastic polymers. *Prog. Polym. Sci.*, 30, 1185, 2005.
6. Subramanian, M.N., Polymer blends and composites, Polymer Blends and Composites: Chemistry and Technology (Polymer Science and Plastics Engineering), Chapter 10, Processing technology, p 237, Scrivener publishing, John Wiley and Sons, 2017.
7. Mills, W. and Robert, A., Tatara, Potential for Reuse of E-Plastics through Processing by Compression Molding. *Challenges*, 7, 17, 2016.
8. Bourry, D. and Favis, B.D., Cocontinuity and phase inversion in HDPE/PS blends: Influence of interfacial modification and elasticity. *J. Polym. Sci. B. Polym. Phys.*, 36, 1889–1899, 1998.
9. Xiu, H., Huang, C., Bai, H., Jiang, J., Chen, F., Deng, H., Wang, K., Zhang, Q., Fu, Q., Improving impact toughness of polylactide/poly(ether)urethane blends via designing the phase morphology assisted by hydrophilic silica nanoparticles. *Polymer*, 55, 1593–1600, 2014.
10. Essawy, H. and El-Nashar, D., The use of montmorillonite as a reinforcing and compatibilizing filler for NBR/SBR rubber blend. *Polym. Test.*, 23, 803–807, 2004.
11. Studies on strength of compression molded waste plastics material – A novel method using additive to improve mechanical strength – Dr. Muralisrinivasan Natamai Subramanian. Unpublished project work.
12. Macosko, C.W., Part I Principles: Constitutive Relations: Elastic Solid, in: *Rheology: Principles, Measurements and Applications*, Wiley-VCH Publishers, New York, NY, USA, 1994.
13. Rader, C.P. and Stockel, R.F., Plastics, Rubber, and Paper Recycling, Chapter 1, Plastics Recycling, ACS Symposium Series; American Chemical Society: Washington, DC, p. 5, 1995.
14. Gahleitner, M., Melt rheology of polyolefins. *Prog. Polym. Sci.*, 26, 895, 2001.
15. Yang, H., Zhan., M., Liu, Y.L., Xian, F.J., Sun, Z.C., Lin, Y., Zhang, X.G., Some advanced plastic processing technologies and their numerical simulation. *J. Mater. Process. Technol.*, 151, 63, 2004.

5

Plastics Waste – Consumer and Industry

Plastics are an attractive manufacturing material due to the fact they can be used in a wide range of applications and exhibit high performance characteristics, including good mechanical strength and toughness, strong dimensional stability and processing advantages. Thermoplastics contribute to major plastic consumption in packaging, construction and consumer goods. Over one-third of the thermoplastic products are for one-time use. The characteristics of plastics provide a barrier to elements in the external environment such as moisture, oxygen, dust, etc. No other conventional material is equivalent to plastics packaging [1]. Improving the quality of life is a continuing global goal for many people, which results in an increased consumption of plastics and produces large amounts of waste. Maximizing the efficiency of energy utilization at every step of the waste disposal process is necessary to achieve sustainable growth, which is an enormous and complex task.

Plastics are a versatile, synthetic long-chain material having lightweight, economical and resistant properties with recyclability. Global production of plastics was 335 million tons in 2016 from 230 million metric tons in 2005. The rapid production of plastics has had numerous societal benefits, making them an indispensable component. There has been a driving annual demand for world production of plastics on a substantial scale [2, 3]. About 50 percent of all plastics are made for single-use applications such as disposable containers and packaging materials [4]. The management of plastics waste optimizes the infrastructure system for managing a given amount and composition of plastics waste [5]. The outcome relies on identifying management techniques which result in the lowest environmental emission and impact.

Muralisrinivasan Natamai Subramanian (ed.) Plastics Waste Management, 2nd Edition (73–96)

5.1 Background

Since the 1950s, 1 billion tons of plastic has been discarded which can persist in the environment for hundreds or even thousands of years [6]; hence, plastics waste has become an ever-increasing and widespread problem over the last few decades. There are many routes for the recycling and recovery of plastics solid waste including; chemical recycling (including pyrolysis, gasification and hydrogenation), through which plastics can be broken down to a feedstock state, and energy recovery *via* the combustion of plastics waste as an alternative fuel source. The recycling and recovery routes of plastics waste [7, 8] contribute to the plastic fraction of the total amount of waste produced, hence there is an urgent need for the proper management of waste plastics [9].

Plastics in many areas of everyday life, from household items to industrial products, can be heated and shaped several times by processing and recycling. The polymer present in plastics varies from about 20–99 percent [10]. Plastics are moldable and made from petroleum and its derivatives. The majority of plastics are resistant to degradation or hydrolytic cleavage. Plastics, such as nylons, polyesters, polycarbonate and polyurethane, are potentially susceptible to degradation by the hydrolytic cleavage reactions due to amide or ester linkage between monomers [11].

The non-biodegradable nature of plastics is due to

- High molecular weight and their distribution in the chain,
- Chemical inertness,
- Presence of additives resistant to oxidation and biodegradation,
- Many plastics are crystalline in nature, which adversely affects biodegradation.

5.2 Plastics Waste

Plastics waste is defined as discarded plastic products or materials which are no longer used in their present form or condition; however, it is a resource which is potentially useful. In reality, certain types of waste may render recovery and reuse impractical, inadvisable or even unfeasible. Recycle management reduces the negative impact that nondegradable plastics waste could have on the environment.

Plastics waste is an inseparable part of human activities due to the day-to-day use of containers, packaging materials, etc. The amount of waste has continued to increase along with population and industrial growth, and the liability of plastics waste hangs over the plastics industry like a blanket. Plastics waste management, waste reduction, and recycling as a source of energy generation, etc., are more environmentally friendly than paper, bioplastic and cotton alternatives. Plastic production and consumption have increased dramatically over recent years, and has become a global environmental concern. This increased amount of plastics waste presents a storage problem [12].

The production of plastics raw material and waste is interdependent, creating both unprecedented opportunities and unprecedented threats for the plastics processing industry. In some cases, sources of waste may generate a single type of plastic waste, but in the majority of cases, the plastic will be a mixture of types.

The increasing annual production rate of polymers is governed by rising demand, which is dependent upon several factors including the economy, while the advancement of polymer technology depends on research. It has become imperative to produce plastics with better properties at reduced cost using improved processes and larger capacity production units. Different groups of plastics exhibit a wide variety of physical properties and even within the same class of plastics, small differences in the preparation method or constituent proportions can result in marked physical differences. The physical properties of plastics are determined by the internal molecular structure as well as the cohesion between long chains or the packing of closed rings. Plastics waste has been recognized as one of the most complex and cost-intensive services.

5.3 Polyolefin

A major source of plastics waste comes from the polyolefin family due to their large consumption as packaging material. Moreover, plastics are hydrophobic in nature, hence easily cleansed of sand and other materials stick with water. However, the oil stick with these materials are difficult to clean. The oil contains oleo-chemicals and therefore it is possible to recycle without any difficulty.

5.4 Polypropylene

When polypropylene is mixed with other plastics, the waste becomes a liability and creates challenging environmental problems. This is due to the fact that when there is a weak interface between waste plastics most physical properties are diminished. The closed-loop recycling of PP is therefore crucial in waste-applications [13].

5.5 Polystyrene

In order to reduce the accumulation of PS in the environment, new recycling/disposal processes should be sought. Among them is the possibility of converting PS into a more easily biodegradable material, though this is a challenge for the materials and microbiological sciences. Although PS is hardly crystalline, its molecules have high molar masses and are non-polar, existing in the glassy state at room temperature (RT); as such, enzymatic attacks are very difficult [14]. Moreover, the phenyl side groups, which are distributed in space in a disordered manner, are biodegraded very slowly [15]. In the case of PS, the weak sites of the polymer chain are the tertiary carbon atoms attached to the phenyl groups, which are vulnerable to attack by free radicals. At these weak sites, a series of chemical reactions can lead to cleavage of the chain and initiate the formation of carbonyl groups [16, 17]; the presence of carbonyl groups, as well as moieties capable of forming free radicals, such as peroxides, accelerates degradation [18, 19].

5.6 Polyvinylchloride

PVC contributes to approximately 80% of the organically bound chlorine found in municipal incinerators and about half of the total chlorine (organic plus inorganic) [20]. During incineration, the chlorine liberated from the incineration process kills the majority of germs present in the municipal waste.

5.7 Bioplastics

Due to living in a plastic age, life without polymers and plastics is difficult. Due to plastic litter and the problems related to their non-biodegradability in the environment, the focus has been on bioplastics as an alternative to commodity plastics in order to minimize the impact of plastics waste on the environment after disposal. The degradable nature of bioplastics depends on the environments they end up in. Bioplastics are nondegradable in marine environments.

Bioplastics are often not as "biodegradable" as they claim to be — it is always a matter of the surrounding environment. Humidity, temperature or concentrations of microorganisms vary in different environments, resulting in different biodegradation rates. The degradability of a polymer is not the only parameter to consider when attempting to replace commodity plastics with biodegradable polymers: potential adverse effects of degradation products on biota also need to be evaluated. In waste management, bioplastics with a controllable lifetime are becoming increasingly important. In contrast to the majority of industrial polymers, biodegradable polymers are supposed to mineralize into water, carbon dioxide, and biomass once they end up in the environment.

With biodegradable plastic products, the use of raw materials and energy in industrial production not only function within a living organism, but with that of the entire natural systems [21]. A natural ecosystem related to biodegradable plastics can be summarized as a food web inhabited by organisms which consume each other or the organic waste of the other organisms. In theory, a natural ecosystem evolves in such a way that no available energy or useful material is wasted.

Developing countries are interested in adopting technologies of higher productivity and increased efficiency, converging in waste reduction [22]. The emerging technologies may bring risks of different wastes. Biodegradable plastics raw material manufacturing and their processing activities have a negative impact on the environment through waste generation, acquiring land, water, air, wood, minerals and metals. The implementation of biodegradable products for sustainability should be specific to the behavior and characteristics of the process.

5.8 Additives and Environment

Additives with metals content such as lead, cadmium, barium, etc., are used to stabilize some plastics during processing. Even though, they metal content is in small amount, within large plastic waste, the metal contents may go up higher and contaminate other environmental sources such as water, ocean, etc. [23, 24].

5.8.1 Heat Stabilizers

The term trace element has been used to define metals present in low concentrations usually being less than 1 mg per kilogram, being classified as essential, non-essential and toxic [25–27]. The elements such as lead, barium and cadmium, etc., heavy metals even at low concentrations in the environment can be extremely toxic to organisms due to the persistence and bioaccumulation in drinking water used for water supply and irrigation purposes with PVC pipes and drain pipes.

The emission cause deleterious environmental impacts at both the regional such as acid rain and global as climate change. Emissions are often used to proxy for the environmental impacts associated with human activities more broadly. Burning waste plastics generates considerable carbon-dioxide which accounts for the emission. Reducing carbon emissions is considered essential if global sustainable development objectives are to be met [28–31].

5.8.2 Plasticizers

The possibility exists for a high level of plasticizers to accumulate in the environment, particularly in different organisms [32, 33]. This requires an investigation into the biogeochemical behavior of plasticizers, with their behavior in natural waters especially being of great importance. It is important from the viewpoint of both ecological protection and environmental risk assessment. The adsorption of these plasticizer particulates and sediments affects other processes of transformation in natural waters, including transport, dissolution, degradation, volatilization and bioaccumulation. These adsorption processes determine the fate of the aquatic system [34]. The toxic properties of these plasticizers are persistent in the environment.

5.8.3 Flame Retardants

Typically, where small household waste is concerned (vegetables, fruits, paper, plastics, glass, bread and carpets), the average concentration of antimony present in dry waste is 5.2 mg/kg [35]. In 1996, a concentration of 7.6 mg/kg of antimony was estimated to be present in household waste in Japan and more than 80% was derived from flame retardant (FR) used in plastics, clothes, curtains and textiles [36]. The antimony concentration in industrial waste is considerably higher than in urban waste, which explains the different antimony concentrations found in the various input waste streams.

One driver that is very clear is the shift of FR demand towards Asia-Pacific and China, and away from Europe and North America. Europe and North America currently account for just over 50% of global FR consumption, but China is poised to challenge for the highest consumption. This situation is largely due to the growing volume of PVC used for lower end plastic items and the increasing requirements for exports to meet overseas flammability requirements. There has been an extensive increase of FR consumption in China and Asia, while demand for non-halogenated organophosphorus FR in North America and Europe has been negatively impacted. The health and environmental scrutiny of FR that has impacted halogenated FR continues to shift towards banning individual classes of products, as opposed to broad-based bans of all FR products. The type of plastic chosen for an application affects the growth and development of individual FR additives.

5.8.4 Compatibilizers

Compatibilizers are compounds of similar chemical structure which can reduce the size of the dispersed phase by lowering the interfacial tension between different phases. This results in a dispersed phase with fine particle resistant to gross separation, and improved overall properties. The waste plastics with compatibilizer are melt blended in a twin screw extruder or an internal mixer to produce a new material.

5.9 Technological Aspects

Plastics technology is pivotal in delivering products and services which fulfil the needs of the population, both now and in the future. Technological input requires resources which are withdrawn from the environment to manufacture products, which society uses and then discards as waste products, i.e., technology requires input which results in products, but finally produces waste [37]. Waste can exist in any form namely, solid, liquid and gaseous, or a combination of the three.

Plastics waste produced by an industrial facility from polymers is the result of processing activities and the reuse of industrial plastics waste has been a common industrial practice for many years. Processing changes, in order to reduce waste and maximize processes, usually require an investigation to determine the costs, alternatives and potential outcomes.

- Reflecting economic growth, developing countries are facing serious plastics waste problems.
- Increase in population and economic growth leads to an increase in waste generation, particularly in urban environments, especially in developing countries.
- Inappropriate waste management causes various environmental problems, such as water, soil and air pollution, resulting in serious health problems.

5.10 Factors Influencing Plastics Waste

The global economy has undergone radical changes which have significant implications for industrial development. The plastic industry has generated mixed waste not only in terms of the quantity of waste, but also its composition. Economic loss from industrial waste must be restricted as discarding plastics waste is detrimental to the economy of industries and society. The planning or development of an industrial waste management system requires knowledge of the quantity of waste which is generated. Any exploitation of the potential for industrial waste recycling relies on detailed knowledge of industrial waste production; however, many problems arise when trying to ascertain the type and quantity of industrial waste. The environment is a key factor that impacts an individual's quality of life and plays a pivotal role in current waste disposal/recycling processes. Industrial waste is a term which refers to all waste produced by industrial operations or derived from manufacturing processes [38].

5.11 Waste Resources

Waste plastics starts from packaging, municipal solid waste, construction and demolition, end-of-life vehicles, medical and biomedical, and waste electrical and electronic equipment.

5.11.1 Domestic Waste

Domestic waste is not only contaminated with wet garbage, but is also far from collection points, which makes sorting difficult. The lightweight plastics or bottles greatly increase operating costs. The plastics to be recycled should operate over a large enough area with an efficient collection system being necessary. The waste from urban areas has proven to be the most difficult and costly aspect of plastics recycling.

Different types of plastics separation may not become economically feasible until the volumes are larger. Mixed plastics could be a good source of potential energy in the case of energy recovery options. In developed countries, the economical separation cost is higher and, moreover,

- Plastics waste by hand sort from municipal waste in bulk by spreading it over a moving conveyor belt proved to be unsuccessful due to the difficulty of finding employees.
- Production, testing and fabrication of plastics devoted mainly to recycling and aimed to export as a source of good raw materials towards production and fabrication.

However, in developing and underdeveloped countries, the cost of the operation with hand sorting is economical due to the availability of employees to work operation and weight to cost separation basis. In the future, even the system of separation of waste from the home starts with biodegradable and non-biodegradable products by keeping two separate buckets. This somehow helps to separate recycled plastics in a better way.

5.11.2 Packaging Waste

The use of plastic materials is constantly increasing. Their use in packaging is significant, consuming plastics at the rate of about 100 million tons annually. They are the largest packaging material, exceeding materials such as metal, glass, paper, paperboard, among others [39]. They extend the shelf life of foods and beverages and provide convenient storage [40–42].

Plastic bags are one of the most commonly used means of packaging because of their versatility, lightweight, flexibility, moisture resistance and strength. Further, plastic bags are mostly relatively inexpensive. It is estimated that on average, 2.7 billion plastic shopping bags are consumed every day worldwide, translating to 1.9 million consumed plastic bags every minute [43].

There is an increase in the demand for food due to an ever-increasing population where plastics packaging has continuous and unabated usage. Plastics packaging waste constitutes an increasing fraction of household waste. Developed countries have executed waste management systems to collect, recover and recycle. However, for a long time now, littering and landfilling of plastics waste has continued in many developing countries [44–46]. This has raised concerns about their increasing occurrence in the soil. Therefore, it poses significant threat to ecosystem. Waste plastics have the non-point sources like runoff during heavy rainfall greatly not only contributed to environmental contamination but also choking cause disaster. Polyethylene terephthalate commonly known as PET are dominate in many of the food packaging. However, polyolefin such as Polyethylene, polypropylene, and polystyrene, dominate as plastics packaging markets and contributes as main waste plastics.

5.11.3 E-Waste

E-waste is the waste material from electronics. It consists of metal and plastics. Apart from quality engineering plastics, in 2012, the generation of this material was at least 3-4 million tons in United States. It is steadily increasing worldwide [47]. This E-waste are being recovered and/or recycled due to its valuable metals such as gold, silver, platinum, palladium, etc., and engineering plastics which cost higher [48].

E-waste also contains additives such as halogen and phosphorus compounds have associated with health concerns. Compression molding into usable end products is the best technique to recycle E-waste. The strength of the end products is suspected due to inclusions unmelt and internal voids from out-gassing. At the same time, an increasing number of states are restricting or banning the landfilling of e-waste. The E-waste are being recycled due to its higher cost. The total amount of E-waste has reached 8.39 million metric tons in Europe per year.

E-waste recycling of plastic materials has attracted attention due to

- Higher cost of virgin plastic materials.
- Equipment population growth beyond the model base years.
- Increasing consumption of electrical and electronic equipment.
- Limited lifespan of electrical and electronic devices [49].

Landfilled or improperly dismantled electronic devices leach and emit extremely hazardous substances, including

- Metals such as lead, cadmium, tin, etc. [50–52]
- Flame retardants such as polyhalogenated biphenyls, bromine flame retardants, etc.
- Plastics with additives such as polyvinylchloride, phthalates, antistatics, etc. [53].

Compression molding into usable end products is the best technique for recycling E-waste. The strength of the end products is thought to be due to the inclusion of unmelted particles and internal voids from outgassing. At the same time, an increasing number of states are restricting or banning the landfilling of e-waste.

5.11.4 Automotive Waste

Acrylonitrile-butadiene-styrene, polypropylene and copolymers, and polyamide are major plastic materials are reprocessed from automotive components from end-life vehicles as foot wears from footwear industries. HIPS is used for manufacturing shoe heel components. The vast majority of scrapped automobiles are processed for recycling starts from shredders and the systems are designed and employed primarily to separate the ferrous and non-ferrous portions of discarded vehicles. The pieces of shredded material are conveyed through a series of separators [54]. Automobile plastics waste are processed for recycling primarily separated from ferrous and non-ferrous portions of discarded vehicles.

5.11.5 Medical Plastics Waste

Thus, medical plastics waste management has been of major concern owing to their potentially high risks to human health and the environment [55, 56]. The HCW generation in developed countries are usually higher than developing countries, probably owing to availability of advanced technologies.

Recycle medical plastics possible to recycle provided plastics packages and wrappers are often generated without any contact with contaminated materials. A correct segregation is essential to ensure adequate procedure of packaging at the time of waste generation [57].

Otherwise, plastics waste generation from the hospital cause an increasing threat to the public and environment. They are hazardous and can be classified as toxic non-biodegradable material. These materials accumulate with time. The accumulation can disturb the surrounding environment. However, a positive method for disposal, such as reuse, would have a beneficial effect. Therefore, there has been increased use of recycled products even in medical applications; however, they require sterilization. As they are recycled materials, there may be a possibility of internal cracks or voids in the products or deterioration during sterilization. Without plastics, what is not stored is not managed; therefore, much expense has been incurred to preserve the medical and food materials for use at a later date. Previously, the predominant constituent of plastics waste in medical waste was polyvinyl chloride (PVC), an inexpensive and common polymer used for both medical and commercial products.

Iatrogenic dioxin pollution can be largely eliminated by avoiding the incineration of PVC. The group of dioxin-like compounds includes all substances with a similar chemical structure and biological effect, which makes particularly troublesome [58, 59].

5.11.6 Agriculture Plastics Waste

Plastic litter could also reduce arable agricultural productivity by reducing aeration in the soil and percolation of water in agricultural fields [60]. Plastic bags littering the environment detract from the natural beauty of a site serving as a focal point for tourism. The excessive use of plastic bags imposes several negative costs to the environment, such as being unsightly, durable litter that is very slow to degrade. The negative effects of plastic bag litter have also been experienced by the main sectors crucial for rural livelihoods, namely agriculture.

5.11.7 Marine Plastics Waste

By 2015, about 6300 metric tons of plastics waste were generated globally. This estimation is based on plastic debris weighing over 260,000 tons floating on the surface of the world's oceans because of indiscriminate waste disposal. Plastic litter is a major pollutant of the marine environment. It is estimated that somewhere between 4.8 and 12.7 metric tons of plastic enter the world's oceans from land per year [61].

Extensive degradation of plastics occurs in nature by photo-, bio-, and thermo-oxidative depolymerization and friction, generating microplastics in the marine environment [62]. Microplastics attract and store toxic compounds such as polychlorinated biphenyls, polybrominated diphenyl ethers, and bisphenols [63]. Microplastic particles have been found in small fish and large mammals [64] in the marine environment.

Polyethylene is very vulnerable to transference by the wind or other natural forces into the sea because of its low density and weight, creating a hazard for humans and wildlife [65, 66] by choking wildlife and distributing harmful organisms or toxic chemicals such as persistent organic pollutants [67]. Therefore, considering these facts and the current depletion of petroleum resources, the recovery of monomers from plastic waste to minimize landfill disposal and fossil fuel consumption is now more important than it has ever been.

5.11.8 Mixed or Contaminated Plastics

Plastics waste disposal with its low biodegradability present in large quantities as mixed waste. Several separation techniques are used to separate individual plastics by density gradient method. The plastics waste is normally with sewage sludge, vegetative residuals, etc. The cleaning and separation consume water and soap. Recycling waste plastics from mixed or contaminated sources are low cost, effective materials. There are only for few technical applications. Mixed or contaminated waste plastics separation and purification will be critical for recycling as raw materials.

In mixed waste plastics, the difficulties are managing, dismantling, identification and separation. To avoid dismantling equipment such as die during processing, it requires scientific knowledge in identification and separation of individual plastics in order to influence recycling.

Plastics waste recovery, with the exception of PET, from the waste stream is to be made compulsory. Plastics waste is subjected to a series of sensors that sort the type and color; further inspection ensures no PVC contamination.

Any waste which is contaminated by soil, or contains dirty plastics, cardboard and so on, cannot be recycled. Small quantities of low-density materials such as low-density PE and PP can be mixed with high-density polyethylene (HDPE); a negligible quantity of PS is not an issue [68]. Figure 5.1 illustrates the separation of plastics from the mixed waste collection.

5.12 Plastics Waste Reduction

Plastics waste prevention appears to require proper bounding of the plastics machinery or processing to be improved. The involvement of industry in the processing and the establishment of proper objectives and targets for improvement. The benefits of plastics waste reduction may take some time to be realized as savings.

Effect of mixed or contaminated waste plastics lead to loss of energy, mechanical properties and burden to the environment. The effects are illustrated in Table 5.1.

Table 5.1 Effect of mixed or contaminated waste plastics.

Mixed or contaminated waste plastics	Effect
Dumping	Total loss of energy and loss of materials
Recycling	Products of lesser value due to loss of mechanical properties.
Thermal recycling in municipal waste (incineration plants or steel works)	Loss of material
Complete degradation of plastics	High dumping costs and legal requirements for recycling.
Unsolved chemical problems	Inertness of the polyolefin and thermodynamic limitations to polymer reversal

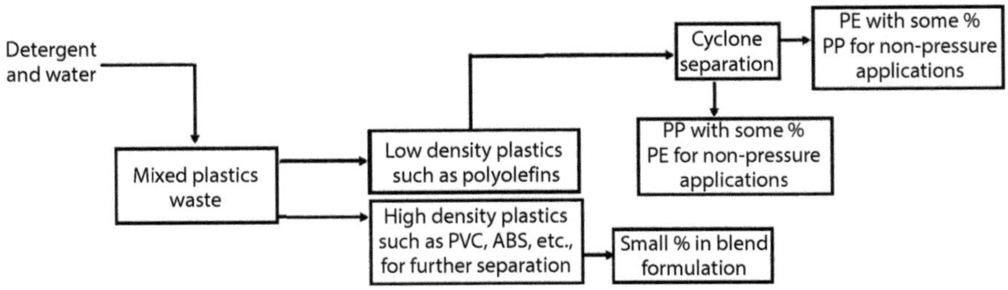

Figure 5.1 Separation of plastics from the mixed waste collection

Plastics waste reduction can be achieved through two primary means

1. Control – reuse, and recycle using processing equipment
2. Prevention – reduce wastage, change or prevent altogether through better housekeeping, material substitution, recycling or process invention [69, 70].

Plastics waste reduction performance improves further reductions in waste often requires more significant changes in processes or even entirely new production technology. Waste reduction to zero will become more technology and capital intensive.

Plastics waste reduction in time period will enhance

- Operating performance
- Financial performance
- Efficient use of assets
- Process improvements
- Cost savings

5.13 Advantages of Waste Prevention

Benefits of waste prevention would appear to be some time lag between the initiation of waste reduction and efforts. Waste prevention provides the following advantages:

- To save the cost of recycling
- Increases productivity and efficiency
- Better utilization of inputs resulting in low raw material.
- Low waste disposal costs.
- Waste prevention strategies offer waste reduction well below the levels.
- May reduce the inventory to realize the cost savings.

In order to integrate environmental concerns in their business operations, integrating waste reduction into existing total quality management (TQM) will be through the use of team involvement. Significant waste reduction appears only possible using continuous improvement methods focused on reuse, recycle and recovery. The logic for waste reduction is analogous to the quality management. Quality prevents defects which is superior to finding and fixing them after the fact [71].

Traditional materials will continue to be replaced by plastic, although the process is expected to slow down. Overall, the output of plastic goods will probably continue to expand slightly and more rapidly than total manufacturing output. Plastics waste processing can be divided into:

- Washing – flushing
- Thermal treatment – solidification

- Reuse – recycling
- Recovery

It is necessary to develop stringent plastics processing and product standards with a preference for permanent solutions that significantly reduce waste volume and which encourage alternatives to land disposal.

5.14 Waste Reduction and Performance

The relationship between waste reduction and industrial performance should be examined.

- Preventing and reducing waste within a short period of time is good for a company's bottom line.
- Industries with minimum levels of waste stand to gain the most.

Thus, reducing waste through recycling would thus appear to be much cheaper than cleaning it up at the end of the pipe [72]. The role of the plastics raw material manufacturing industry forms important aspects of the international dimension of plastics waste and the environment.

The plastics to be recycled should operate over a large enough area with an efficient collection system. In the production of plastics raw materials, the waste material is added to a percent and comes to the market as virgin material. Sometimes the plastics raw material manufacturer disposes of waste material at an economical cost compared to virgin material, and further recycling of the material becomes very difficult due to weak properties.

5.15 Recovery of Plastics

The recovery of plastic materials from industrial processing is both environmentally necessary and economically feasible. Developing practical methods to enable the utilization of plastics waste after processing is necessary for two reasons: the diverse applications of recycled products for a number of consumer and industrial applications, and to ensure the waste generated is recycled or has an alternative use.

The development of the plastic industry has generated a large quantity of solid waste, especially non-degradable waste. Plastics waste from industrial activities contains additives; hence, the practice of segregating plastics waste reduces potential environmental, health and safety impacts, and ensures recycling facilities are cost-effective. The essential step involves cleaning and recycling. In many countries, the treatment infrastructure is inadequate. The main practices for plastics waste are currently:

- The collection of plastics waste from industries and disposal at domestic landfill sites without any segregation.
- Recycling at waste generation sites.
- Disposal within the waste generation stream.
- Reuse and informal recycling.

5.16 Remarks

Key challenges with various factors drive the demand for the recovery of material from industrial waste. The economic value of the plastics industry relies on virgin and recycled material, the technology of using virgin material, the substitution of recycled material for virgin material and the market structure. The lack of record-keeping within industrial sectors presents a major challenge for the verification of the actual amounts of materials that are eventually recycled; knowledge of the energy saved as a result of the reuse of recycled materials is a key factor in the utilization of industrial waste.

Due to the rapidly increasing growth of the plastics industry, a large quantity of plastic products become obsolete every year; therefore, outdated plastic products, including end-of-life plastic products, are the most rapidly growing plastics waste problem worldwide [73]. As a result, many plastic materials, such as polyvinyl chloride, acrylonitrile-butadiene-styrene and so on, along with heavy metals and flame retardants, create serious pollution problems during the recycling and disposal process.

References

1. Yam, K.L., Takhistov, P.T., Miltz, J., Intelligent packaging: Concepts and applications. *J. Food Sci.*, 70, R1–R10, 2005.
2. Andrady, A.L. and Neal, M.A., Applications and societal benefits of plastics. *Philos. Trans. R. Soc. London, Ser. B*, 364, 1977–1984, 2009.
3. Thompson, R. C., Moore, C. J., vom Saal, F. S., Swan, S. H., Plastics, the environment and human health: current consensus and future trends. *Philos. Trans. R. Soc. London, Ser. B*, 364, 2153–2166, 2009.
4. Jambeck, J.R., Geyer, R., Wilcox, C., Siegler, T.R., Perryman, M., Andrady, A., Narayan, R., Lavender Law, K., Plastic waste inputs from land into the ocean. *Science*, 347, 6223, 768–771, 2015.
5. Wang, H. and Nie, Y., Municipal Solid Waste Characteristics and Management in China. *J. Air Waste Manage. Assoc.*, 51, 250, 2001.
6. Weisman, A., in: Part II, Chapter 9, Polymers are forever, *The World Without Us*, Thomas Dunne Books, St. Martin's Press, New York, NY, USA, 2007.
7. Subramanian, P.M., https://www.worldcat.org/title/plastics-recycling-and-waste-management-in-the-us/oclc/4932897273&referer=brief_results" Plastics recycling and waste management in the US, *Resour. Conserv. Recycl.*, 28, 253, 2000.
8. Thanh, N.P., Matsui, Y., Fujiwara, T., Assessment of plastic waste generation and its potential recycling of household solid waste in Can Tho City, Vietnam. *Environ. Monit. Assess.*, 175, 23, 2011.
9. Dewulf, J.P. and Van Langehove, H.R., Quantitative Assessment of Solid Waste Treatment Systems in the Industrial Ecology Perspective by Exergy Analysis, *Environ. Sci. Technol.*, 36, 5, 1130-1135, 2002.
10. Sethi, B., Chapter 3 Methods of Recycling, Recycling of Polymers Methods, in: *Characterization and Applications*, Francis R. (ed.), Wiley-VCH, p 62–63, 2016.
11. Muller, R.J. and Deckwer, W.D.K.I., Biodegradation of polyesteres containing aromatic constituents. *J. Biotechnol.*, 86, 87–95, 2001.
12. Mączka, T., Śliwka, E., Wnukowski, M., Plasma gasification of waste plastics. *J. Ecol. Eng.*, 14, 1, 33–39, 2013.
13. Lennart, Y.L., Materials selection and design for development of sustainable products. *J. Mater. Des.*, 28, 466–79, 2007.
14. Motta, O., Proto, A., De Carlo, F., Santoro, E., Brunetti, L., Capunzo, M., Utilization of chemically oxidized polystyrene as co-substrate by filamentous fungi. *Int. J. Hyg. Environ. Health*, 212, 1, 61, 2009.

15. Atlas, R.M. and Bartha, R., Biogeochemical cycling: Nitrogen, sulfur, phosphorus, iron, and other elements. *Microbial Ecology: Fundamentals and Applications*. Menlo Park, CA, The Benjamin/ Cummings Publishing Company, Inc.: 333–360, 1987.
16. Severini, F., Gallo, R., Ipsale, S., Environmental degradation of stabilized LDPE. Later stages. *Polym. Degrad. Stab.*, 17, 1, 57, 1987.
17. Niessner, N. and Gausepohl, H., Chapter 2, Polystyrene and Styrene Copolymers - An Overview, Scheirs, J. and Priddy, D.B. (Eds.), *Modern Styrenic Polymers:Polystyrenes and Styrenic Copolymers*, p. 703, John Wiley & Sons, Chichester, UK, pp. 25–41, 2003.
18. Botelho, G., Queiros, A., Machado, A., Frangiosa, P., Ferreira, J., Enhancement of the thermooxidative degradability of polystyrene by chemical modification. *Polym. Degrad. Stab.*, 86, 3, 493, 2004.
19. Meekum U. and Kenharaj, R., Comparative study of polystyrene foam degradation in the open-air and artificial weathering. *Arabian J. Sci. Eng.*, 27, 1C, 25–32, 2002.
20. Thornton, J., McCally, M., Orris, P., Weinberg, J., Dioxin Prevention and Medical Waste Incinerators. *Public Health Rep.*, 111, 4, 1996.
21. Frosch, R.A. and Gallopoulos, N.E., Strategies for manufacturing. *Sci. Am.*, 261(3): 144–152, 1989.
22. Vaccari, M., Tudor, T., Perteghella, A., Costs associated with the management of waste from healthcare facilities: An analysis at national and site level. *Waste Manage. Res.*, 36, 39–47, 2018.
23. Yachigo, S.i., Chapter 9, Synergistic Stabilization of Polymers, *Handbook of polymer degradation,* S. Halim Hamid (ed.), Mohamed B. Amin, Ali G. Maadhah, p305, Marcel Decker, New York, 1992.
24. Williams, P.T. and Williams, E.A., Interaction of plastics in mixed plastics Pyrolysis. *Energy Fuels*, 13, 188–196, 1999.
25. Batzias, A.F. and Siontorou, C.G. A new scheme for biomonitoring heavy metal concentrations in semi-natural wetlands, J., *Hazard. Mater.*, 158, 340–358, 2008.
26. Fernandes, C., Fontainhas-Fernandes, A., Peixoto, F., Salgado, M.A., Bioaccumulation of heavy metals in Liza saliens from the Esmoriz–Paramos coastal lagoon, Portugal. *Ecotoxicol. Environ. Saf.*, 66, 426–431, 2007.
27. Sekhavatjou, M.S., Alhashemi, A.H., Rostami, A., Comparison of Trace Element Concentrations in Ambient Air of Industrial and Residential Areas in Tehran City. *Biol. Trace Elem. Res.*, 143, 1413–1423, 2011.
28. Shan, Y., Guan, D., Hubacek, K., Zheng, B., Davis, S.J., Jia, L., Liu, J., Liu, Z., Fromer, N., Mi, Z., Meng, J., Deng, X., Li, Y., Lin, J., Schroeder,

H., Weisz, H., Schellnhuber, H.J., City-level climate change mitigation in China. *Sci. Adv.*, 4, 6, eaaq0390, 2018.
29. Mi, Z. *et al.*, Carbon emissions of cities from a consumption-based perspective. *Appl. Energy*, 235, 509–18, 2019.
30. Fry, J. *et al.*, Assessing carbon footprints of cities under limited information. *J. Clean Prod.*, 176, 1254–70, 2018.
31. Chen, S., Xu, B., Chen, B., Unfolding the interplay between carbon flows and socioeconomic development in a city: What can network analysis offer? *Appl. Energy*, 211, 403–12, 2018.
32. Xu, X.R., Li, H.B., Gu, J.D., Li, X.Y., Kinetics of n-butyl benzyl phthalate degradation by a pure bacterial culture from the mangrove sediment. *J. Hazard. Mater.*, 140, 194–199, 2007.
33. Xu, X.R. and Li, X.Y., Adsorption behaviour of dibutyl phthalate on marine sediments. *Marine Pollution Bulletin*, 57, 403–408, 2008.
34. Coulson S.H., Polymer Types and Additives, in *Plasticizers, Stabilizers and Fillers*, P.D. Ritchie, S.W. Critchley, A. Hill, (Eds.), Iliffe Books Ltd., London, UK, pp. 9–38, 1972.
35. Nakamura, K., Kinoshita, S. and Takatsuki, H., The origin and behavior of lead, cadmium and antimony in MSW incinerator. *Waste Manage.,* 16, 509–517, 1996.
36. Gershon, A.L., Gyger, L.S., Jr., Bruck, H.A., Gupta, S.K., Thermoplastic Polymer Shrinkage in Emerging Molding Processes. *Exp. Mech.*, 48, 789, 2008.
37. Abduli, M.A., Industrial waste management in Tehran. *Environ. Int.*, 22, 335, 1996.
38. Hogland, W. and Stenis, J., Assessment and system analysis of industrial waste management. *Waste Manage.*, 20, 537, 2000.
39. Cooper, T.A., Chapter 4, Developments in plastic materials and recycling systems for packaging food, beverages and other fast-moving consumer goods, in: *Trends in Packaging of Food, Beverage and other Fast-moving Consumer Goods (FMCG)*, p. 58, Woodhead Publishing Limited, Cambridge, U.K. 2013.
40. Hutton T., Food Packaging: *An introduction. Key topics in food science and technology*. Chipping Campden, Gloucestershire, UK: Campden and Chorleywood Food Research Association Group, 7, 108, 2003.
41. Kirwan, M.J. and Strawbridge, J.W., Chapter 7 Plastics in food packaging, Coles, R., McDowell, D., Kirwan, M.J. (Eds.), *Food packaging technology*, CRC Press, Boca Raton, 2003.
42. Al-Salem, S.M., Lettieri, P., Baeyens, J., Recycling and recovery routes of plastic solid waste (PSW): A review. *Waste Manage.*, 29, 2625–2643, 2009.

43. Madigele, P.K., Mogomotsi, G.E., Kolobe, M., Consumer willingness to pay for plastic bags levy and willingness to accept eco-friendly alternatives in Botswana. *Chin. J. Popul. Resour. Environ.*, 15, 255–261, 2017.
44. Agamuthu, P. and Victor, D., Policy trends of extended producer responsibility in Malaysia. *Waste Manage. Res.*, 29, 945–953, 2011.
45. Zurbrügg, C., Caniato, M., Vaccari, M., How Assessment Methods Can Support Solid Waste Management in Developing Countries—A Critical Review. *Sustainability*, 6, 2, 545–570, 2014.
46. Salau, O., Osho, S., Sen, L., Osho, G., Salau, M., Urban Sustainability and the Economic Impact of Implementing a Structured Waste Management System: A Comparative Analysis of Municipal Waste Management Practices Developing Countries. *Int. J. Reg. Dev.*, 4, 1, 1–13, 2017.
47. Buekens, A. and Yang, J., Recycling of WEEE plastics: A review. *J. Mater. Cycl. Waste Manag.*, 16, 415–434, 2014.
48. Wang, R.X. and Xu, Z.M., Recycling of non-metallic fractions from waste electrical and electronic equipment (WEEE): A review. *Waste Manage.*, 34, 1455–1469, 2014.
49. Babu, B.R., Parande, A.K., Basha, C.A., Electrical and electronic waste: A global environmental problem. *Waste Manage. Res.*, 25, 307–318, 2007.
50. Gu, W., Bai, J., Yao, H., Zhao, J., Zhuang, X., Huang, Q., Zhang, C., Wang, J.W., Heavy metals in soil at a waste electrical and electronic equipment processing area in China. *Waste Manage. Res.*, 35, 1183–1191, 2017.
51. Chen, H., Tenga, Y., Lu, S., Wang Y., Wang, J., Contamination features and health risk of soil heavy metals in China. *Science of the Total Environment,* 512–513, 143-153, 2015.
52. Dimitrakakis, E., Janz, A., Bilitewski, B., Gidarakos, E., Small WEEE: Determining recyclables and hazardous substances in plastics. *J. Hazard. Mater.*, 161, 913–919, 2009.
53. Morf, S., Tremp, J., Gloor, R. Schuppisser, F., Stengele, M., Taverna, R., Metals, non-metals and PCB in electrical and electronic waste - Actual levels in Switzerland. *Waste Manage.*, 27, 1306–1316, 2007.
54. Vandermerwe, S. and Oliffe, M.D., Corporate challenges for an age of reconsumption. *Columbia J. World Bus.*, 26, 3, 6, 1991.
55. Oyekale, A.S. and Oyekale, T.O., Healthcare waste management practices and safety indicators in Nigeria. *BMC Public Health*, 17, 740, 2017.

56. Bazrafshan, E., Mohammadi, L., Mostafapour, F. K., Moghaddam, A. A., Dental solid waste characterization and management in Iran: A case study of Sistan and Baluchestan Province. *Waste Manage. Res.*, 32, 157–164, 2014.
57. Ali, M., Wang, W., Chaudhry, N., Yong, G., Hospital waste management in developing countries: A mini review. *Waste Manage. Res.*, 35, 6, 581–592, 2017.
58. Cheng, Y.W., Sung, F.C., Yang, Y., Lo, Y.H., Chung Y.T., Li, K-C., Medical waste production at hospitals and associated factors. *Waste Manage.*, 29, 440–444, 2009.
59. International Committee of the Red Cross, Medical Waste Management, Chapter 2, Definition and Description of "Medical Waste" ICRC, 19, avenue de la Paix, 1202 Geneva, Switzerland, pp. 11–14, 2011.
60. Njeru, J., The urban political economy of plastic bag waste problem in Nairobi, Kenya. *Geoforum*, 37, 1046–1058, 2006.
61. Jambeck, J.R., Geyer, R., Wilcox, C., Siegler, T.R., Perryman, M., Andrady, A., Narayan, R., Lavender Law, K., Plastic waste inputs from land into the ocean. *Science*, 347, 6223, 768– 771, 2015.
62. Browne, M.A., Crump, P., Niven, S.J., Teuten, E., Tonkin, A., Galloway, T., Thompson, R., Accumulation of microplastic on shorelines worldwide: Sources and sinks. *Environ. Sci. Technol.*, 45, 21, 9175–9179, 2011.
63. Bouwmeester, H., Hollman, P.C., Peters, R.J., Potential health impact of environmentally released micro-and nanoplastics in the human food production chain: Experiences from nanotoxicology. *Environ. Sci. Technol.*, 49, 8932–8947, 2015.
64. Boerger, C.M., Lattin, G.L., Moore, S.L., Moore, C.J., Plastic ingestion by planktivorous fishes in the North Pacific Central Gyre. *Mar. Pollut. Bull.*, 60, 12, 2275–2278, 2010.
65. Hopewell, J., Dvorak, R., Kosior, E., Plastics recycling: Challenges and opportunities. *Philos. Trans. R. Soc. London, Ser. B*, 364, 2115–2126, 2009.
66. Verma, R., Vinoda, K., Papireddy, M., Gowda, A., Toxic Pollutants from Plastic Waste- A Review. *Procedia Environ. Sci.*, 35, 701–708, 2016.
67. Barnes, D., Galgani, F., Thompson, R., Barlaz, M., Accumulation and fragmentation of plastic debris in global environments. *Philos. Trans. R. Soc. London, Ser. B*, 364, 1985–1998, 2009.
68. Steuteville, R., MSW composting at the crossroads. *Biocycle*, 36, 44–53, 1995.

69. Frosch, R. and Gallopoulos, N., Strategies for manufacturing. *Sci. Am.*, 261, 144–152, 1989.
70. Oakley, B.T., Chapter 25, Total Quality Product Design - How to Integrate Environmental Criteria into the Product Realization Process, Environmental TQM, 2nd Ed., John T. Willig, Ed.,New York: McGraw-Hill, 1994.
71. Imai, M., *Kaizen*, Random House Business Division, New York, 1986.
72. Toyota Motor Corporation, Toyota: A History of the First 50 Years, 39–40. Toyota City, Japan, Dai Nippon, 1988.
73. Halluite, J., Linton, J.D., Yeomans, J.S., Yoogalingam, R., The challenge of hazardous waste management in a sustainable environment: insights from electronic recovery laws. *Environ. Manage.*, 12, 31, 2005.

6
Plastics Waste Management

Plastic materials are universally used in all societies. The advantages of plastic have contributed decisively to improve the lifestyle of a society. Plastics generate a wide range of environmental impacts from polymer manufacturing, processing and products. The main problem is the end-of-life of plastics, which decides the best plastics waste management. Therefore, waste plastics is of critical importance.

Both the use of plastics and the subsequent generation of plastics waste are increasing rapidly worldwide. Plastics waste refers to the amount of materials and products generated prior to recycling, landfilling and incineration, which is governed by the density and living standard of the population [1], income level [2], economic growth, consumption pattern, and waste collection and recycling framework.

6.1 Principles

The theoretical principles of plastics waste management describe the ideal plastics processing. Plastics waste management is used to achieve minimum waste in processing. The management of plastics waste presents enormous challenges. It will take a great deal of time and effort to bring about the necessary fundamental changes to the patterns of production and consumption.

Muralisrinivasan Natamai Subramanian (ed.) Plastics Waste Management, 2nd Edition (97–114) © 2019 Scrivener Publishing LLC

6.2 Objective

The objective of plastics waste management is

- To assess the environmental performance and potential impacts of the system under examination.
- To eliminate waste with energy conserving methods, environmentally friendly products, and a recycling network to collect and reuse plastics waste.
- To cut waste at each manufacturing facility, technical center, and parts distribution center, autonomous waste reduction facilities are established to implement management plans and eliminate waste generation.
- To integrate waste regulations and adopt an integrated management system for the reduction of waste.

6.3 Requirements

Plastics waste management is essentially the planning and utilization of waste in a better manner. It allows a problem or issue to be considered in terms of a pattern of behavior over time and results in considering the current situation in the context of timescale; the strategy should thus have a historical segment, a current state and one or more future paths [3].

Plastics waste management currently requires the simultaneous consideration of seven factors, which include:

1. Collection
2. Segregation
3. Scientific thinking
4. Waste quantity estimation
5. Qualifying the waste
6. Separation and cleaning
7. Recycling into product

In order to optimize waste management, the seven factors listed above remain the fundamental elements to be considered.

6.4 Management Concept

Plastics waste management involves the process of disposal, i.e., managing the transformation of waste into products or services. The concept of plastics waste management extends beyond the processing function of the plastics industry. Therefore, plastics waste management needs

- To associate the concept with the processing or recycling behavior of the material. An abundance of plastics waste is generated between processing and consumption, between reuse and recovery and between the environment and human business activity. A reduction of plastics waste results from recycling. The environment and plastics waste could be considered patterns of recycling behavior, such as the fossil fuel produced to meet energy demands.
- To effectively match material production with reduced waste generation. Plastics waste itself is replete with adaptive internal/external boundaries from products and consumers.
- To include recycling management as part of processing. The end-of-life phase of plastics waste is important from an environmental point of view and there is economical potential in improving material efficiency by climbing the waste hierarchy. Even effective and environmentally aware industries have opportunities to improve plastics waste management.
- To acknowledge that the waste recycling process operating over the long term likely indicates the organizational goal, as industries most clearly rely on the environment for necessary input.

6.5 Waste Collection

The basic operation of plastics waste management involves: waste collection, transportation, storage and treatment. It requires each step of the operation requires expertise and equipment. Although these functions remain similar, as environmental management progresses, the roles and applications of products may change; therefore, an ideal waste management system is function oriented, not product oriented [4–8].

Plastics waste management has evolved from simple to complex integrated systems. In the systems, the plastics waste is not only transported and disposed of, but also reused and recycled to produce valuable products. Plastics waste collection, pretreatment, transport, reuse and recycling of wastes has to be synchronized to have successful plastics waste management. Recycling is built to optimize existing systems, as well as design and implement management systems.

6.6 Separation and Cleaning

Separating plastics requires little scientific research and deserve more attention. Plastics waste require sustainable and effective separation, sorting and cleaning. Collection and processing leaves options open for separation and recovery. Plastics waste disposal with low biodegradability is present in large quantities as mixed waste.

From municipal solid waste, plastics waste is normally with sewage sludge, vegetative residuals, etc. The cleaning and separation consume water and soap. In mixed waste plastics, the difficulties are management, dismantling, identification and separation. Mixed or contaminated waste plastics separation and purification will be critical for recycling as raw materials. It requires scientific knowledge in identification and separation of individual plastics in order to influence recycling and to avoid dismantling equipment such as die during processing. Therefore, the separation of plastics waste from home and industry is essential. Recycling at source after separation causes less pollution. It requires less direct energy consumption and lower capital costs in comparison to mechanical separation of mixed wastes.

Several separation techniques are used to separate individual plastics; however, the separation methods mainly include electrostatic sorting, floatation, etc. Vibration and pneumatic sorting are relatively lower cost and without any environmental pollution of air, water, electromagnetism. Due to density variations of plastics film waste are narrow and uneven wind force [9–12].

6.7 Scientific Thinking

The traditional waste collection and disposal pathway is prevalent, albeit with numerous shortcomings. Throughout the world, the objective of modern waste management has shifted from the disposal of wastes to supplying the economy with secondary raw materials and recovered energy, for which reason waste to energy has become an innovation of interest [13].

6.8 Outcome

The plastics waste operation looks at the structure or 'physics' of relationships, at how one variable affects another; not just that they affect each other.

- Plastics waste management helps to recognize the concept of interdependence; that generally within a system, there is a web of relationships [14].
- Plastics waste management needs to develop a reliable method of estimating waste composition. Certain variations can be observed, which can be attributed to specific social activities.

6.9 Effective Management

The successful planning of plastics waste management program relies on the availability of reliable information regarding the quantity and type of material being generated, and an understanding of how much of that material the collection program can expect to prevent or capture. Statistical data should be used to establish a baseline for monitoring progress in achieving waste management objectives.

The effective plastics waste management can be achieved by taking into account:

- Need to estimate material recovery potential
- Identifying sources of component generation
- Facilitating the design of processing equipment
- Estimating the physical, chemical and thermal properties of the waste and
- Maintaining compliance with national laws and European directives.

The composition of generated waste is extremely variable as a consequence of seasonal, lifestyle, demographic, geographic and legislative factors. This variability makes defining and measuring the composition of waste more difficult and at the same time essential. It is impossible to reduce solid waste to zero level of pollution. As long as society lives in a culture of consumption, the plastics waste problem will reach crisis proportions and pose a challenge to public policy-making systems and governance [15].

6.10 Dynamic Thinking

Plastics waste management builds on dynamic thinking, i.e., seeing the bigger picture. It requires the ability to 'rise above' functional silos and 'view' the recycling process that links the recycled products. The plastics waste process enables the determination of plausible explanations to be identified, i.e., relationships that are not under the control of a decision within a system should be eliminated from consideration. Essentially, this perspective means viewing plastics waste as a component of solid waste and as such under the control of recycling decisions.

The plastics waste process helps to identify closed-loop structures and the process of recycling plastics waste does not run in just one direction, but an 'effect' usually feeds back to change one or more of the 'causes', and that the 'causes' themselves affect each other. It is an important part of the recycling process not to prioritize 'causes' as being most or least important, but rather to understand how waste composition and quantity may change over time.

6.11 Multi-Phase Approach

Plastics waste management is a multi-phase approach to management and recycling with a setup phase of identifying objectives and other decision elements such as reuse or recovery. An iterative phase uses these elements in an ongoing production cycle of technology having processing and management impacts. Industrial phase involves the periodic reconsideration of the decision elements. Industrial learning is complementary to that of processing, including uncertainty waste, propagation of production change and consequences of change over time. Processing issues include ways to recognize when the decision elements should be revised, which elements should be utilized, and how alternatives can be identified and incorporated based on experience and management performance.

6.12 Significance

In plastics waste management, polymer production transcends the level of the individual industry and encompasses the whole product chain from cradle to grave. Consequently, in the plastics processing industry, waste management is not a familiar concept outside the polymer industry. Although life cycle management is used to express the basic philosophy behind plastics waste management, plastic products in processing depend on the implementation of product improvements between production and reuse. However, the general principles of plastics waste management are not clearly formulated to motivate industries to adopt the concept. Insight is needed into the specific activities that industries should carry out to implement plastics waste management.

Plastics waste management is becoming increasingly important at the industrial level. This is due, in part, to environmental improvements and the integrated framework of recycle performance. The management is divided into four categories, namely:

1. Production planning
2. Quality management systems
3. Processing and products performance
4. Waste reduction and recycling

6.13 Progressive Management Characteristics

The world is rapidly moving towards a future with plastics. The majority of available products are almost all made with plastics. The plastics waste management systems are ubiquitously deployed to support society. Intelligent waste plastics management promises a new era of economic gains and improved quality of life. However, to realize these benefits, the life cycle and recycling must perform flawlessly. Perfecting the life cycle of plastics waste management underpinning plastics waste presents an exciting challenge for the next century of plastics in society. Progressive management will focus on the following major areas to control waste plastics in the future:

(1) Control of plastics raw material manufacturing
(2) Plastics processing technologies and their efficiency
(3) Life cycle assessment of plastics
(4) Enhanced recycle, reuse and recovery techniques
(5) Technology to process contaminated or mix plastics waste
(6) Recycling waste as by-products

In order to address the challenges, it is necessary to implement far-reaching recovery programs. As a result, more research and development will necessary. However, moving to totally virtual raw material mode of divestments and manufacturing units impact the plastics processing industry. Therefore, opportunities are to be created to encourage recycling as new business prospects. There is widespread awareness concerning plastics waste. The outlook for life cycle assessment in the immediate future is, at best, uncertain. The plastics waste management systems, along with the statutory need to recycle or recover, will also increasingly become raw material production. Because the economic advantage of recycling materials with the present plastics waste has not yet been fulfilled due to the supply and production of plastics raw material, there will be further depletion of renewable resources. The necessary solution is to shut down the production of plastics raw material production and encourage the recycling unit.

6.14 Risks in Plastics Waste Management

Plastics waste management addresses uncertainty about the waste influencing production dynamics, as well as the elements of decision making itself. Management is used to reduce both kinds of uncertainty in known procedures to address waste uncertainty and production progress. The decision elements which complement the framework for reducing waste uncertainties are:

- Waste cleanup
- Process risks
- Financial availability and risks – Perhaps the decision to innovate the plastics waste remediation technologies is more of a financial decision than a technological one. Innovating industries may simply be financial risk-takers. Therefore, technology appropriation may not rank as highly as in industries. Industries will focus on the adaptation of existing technologies rather than attempting radical innovation programs, which will remain low to minimize the financial risks.
- Uncertainty of waste generation
- Engineering difficulties
- Risks of liability

6.15 Factors – Affect, Suffer, and Influence

In plastics waste management, learning opportunities, such as side-loading containers collected by one-man compactor trucks, automated waste sorting lines similar to mining technology, and enhanced operations of fluidized bed incineration, are some of the ideal ways to start accelerated disposal techniques [16].

Key factors that affect plastics waste management are [17, 18]:

- Technology
- Environmental systems
- Financial systems
- Socio-cultural systems
- Institutional and legal systems

6.16 Operational Problems

Plastics waste management suffers from the following [19, 20]:

- Lack of consistency in plastics waste management
- Lack of completion due to corruption, complex legislation, new elections to replace project staff, lack of funds, etc.
- No dedication and motivation to complete the project
- New political structure with less qualified personnel and internal decision makers
- Project initiation without any motivation
- Difficulty in formal finalization

Different factors influencing plastics waste management are:

- Legislation and regulation
- Institutional and technology development
- Regional and international laws
- Socio-economic factors

6.17 Sustainability and Symbolic Management

Plastics waste management as it relates to sustainability involves a real change in processes or practices such as the improvement in quality and environmental information in order to respond to consumers' demands [21]. This may become a symbolic managerial strategy to enhance industrial status [22, 23], such as the use of heavy-metal free additive in plastic products to sustainably manage their hazardous nature [24].

Sustainability provide information about economic, social and environmental impacts and also environmental accountability [25]. The information may be a driver to

- Build trust,
- Improve processes and systems,
- Progress on the industries,

- Reduce compliance costs, and
- Create competitive advantages.

6.18 Environmental Conservation

Plastics waste management is an important facet of environmental conservation and encompasses planning, organization, administration, financial and legal aspects of various activities and is required to be environmentally compatible, and employs the principles of economy, aesthetics, energy and conservation [26]. Unscientific waste management aggravates different types of environmental pollution, such as air and water, including groundwater and soil quality, which can impact public health as waste is a breeding ground for various diseases [27].

Gross productivity of urban waste services covers only approximately one-third of the overall cost [28]. Correct planning for the primary and secondary collection, and subsequent disposal, of plastics waste has become crucial to ensure a clean, healthy and pathogen-free urban environment.

The opportunity provided by integration of plastics waste management improves the efficiency and performance of recycling, reuse and recovery of plastics waste. It enables dynamic, adaptable and optimized operations for technological improvements in waste plastics conversion. The plastics waste conversion results from technological advances induced by revolutionary impacts on the economy.

The efficiency improvements result in the introduction of a reduction in the plastics waste by optimization of production, reuse and recycle. The plastics waste management can adjust power consumption in response to the cost. Reuse of plastics waste could delay the need to add capacity to the production. Reduction of waste results in further improvement when using virgin materials.

6.19 Decision-Making Process

The recycling of plastics waste has increasingly been accepted as a response to a complex situation and the default understanding of this situation has not yet led to adequate action. However, despite the accepted value and ease of recycling plastics waste, most individuals have a great deal of difficulty in systemically recycling waste. In addition, it has been found that plastics waste processes are insensitive to feedback and underestimate time lags between actions and responses. This insensitivity to feedback reflects a failure of the decision-making process to correctly assess the nature and significance of the solid waste management system, particularly the link between the decision and the environment. Plastics waste management has been astonished by how hard it is for firms to repeat their success when technology or markets change. The failure of plastics waste management can sometimes be ascribed to managerial myopia or organization lethargy, or to insufficient resources or expertise. Many developed countries have great competitive awareness, aggressively investing in recycling and technologies and astutely listening to industrialists. Therefore, plastics waste management involves the manner in which waste is transformed into materials, innovation, technology, information, products or services. Here, innovation refers to a change in technology and technology refers to how an organization transforms labor, capital and materials [29].

6.20 Integrated Plastics Waste Management

Integrated plastics waste management and strategies help to reduce waste plastics. Major portions of plastics waste are under reprocessed or recycled. Plastics waste generation is an inevitable consequence of production and consumption activities related to the level of income of a population and degree of urbanization [30]. The quantity and characteristics of plastics waste are the two major factors which govern the design of an efficient, cost-effective and environmentally compatible waste management system. Integrated systems involve a number of management techniques which maximize the recovery of resources from the waste.

Rapidly expanding urban activities, including commercial and industrial development, along with improved lifestyles due to socio-economic development, have resulted in an enormous quantity of solid waste. The world is facing a crisis in plastics waste management as there is no suitable permanent site to dump the waste or an integrated scheme for its sustainable management. In addition, information on the status of plastics waste management is insufficient. The practices and sources of segregation, primary and secondary collection, including composition and characterization, and disposal practices are summarized below. Effective plastics waste management requires knowledge of factors such as:

- The source of waste generation.
- The existing status of waste collection, transportation and segregation of waste.
- Waste characteristics and disposal activities.

Therefore, integrated plastics waste management is generally focused on

- Incremental innovations of accepted technologies such as thermal treatment in the case of contaminated plastics.
- An integral part of industrial philosophy from the top levels of management.

Plastics waste management was designed for the domestic collection of waste. Industrial metabolism and ecology have motivated the identification of waste that can be recovered, upgraded, and reused as alternative raw materials, thereby closing the cycle of materials [31, 32].

6.21 Assignments

Plastics waste management is rooted in cognitive processes. The concept of solid waste as a paradigm requires considering the process holistically; however, translating the plastics waste paradigm into 'measurable' elements has remained a research challenge. The management of plastics waste seeks to identify causality, i.e., determine how the waste is generated. Generally, society has a tendency to think in terms of 'correlation' or 'influence.'

Plastics waste management is related not only to reuse, recycling, energy recovery and landfilling, but also prevention and/or minimization of waste [33]. Technological changes in plastics processing have led to products being thrown away as or before performance declines.

Plastics waste management is a relatively new area of study from processing oriented to questions about the meaning of waste from industries and its value. It is an activity related to developments that may demonstrate very different responses to waste issues.

- Empirically based on the processing issues of the elements of waste with scrutinizing links with raw materials and additives. However, the values and beliefs processing issues and trends play a significant part in determining the manner in which the industry operates.
- Industry has to some extent omitted the fact that waste may demonstrate very different responses to, for instance, environmental issues.

A clear understanding of plastics waste management requires an organization to be process-centric and involves many interlocking processes – some functional and some cross-functional, and the latter will be built upon the foundation of strong waste management skills and processes. The economy is now entering its fourth stage and is determined by many factors including limited fossil fuel resources and plastics waste management. Many industries are in this post-industrial economic stage due to their handling of waste management skills, and competitive advantages will result from the most effective processes. A process-centric management focus and organization will replace a task-centric perspective and management style [34].

6.22 Advantages

- Plastics waste management has been reaping profits in the case of recycling with a strong interest in their growth.

- Plastics waste management is becoming increasingly accepted as a key feature of a business strategy.
- Plastics waste management is a shared world view consisting of beliefs, assumptions, and values of a plastics waste regarding the relationship of its activities towards processing.

6.23 Shortcomings

- Characteristics and gaps in the existing process and environmental improvement tools and principles.
- Material waste and inefficiencies in the handling of materials and waste recycling implementation.
- Plastics waste recycling by analyzing the existing body of tools for processing and environmental improvement.
- Lack of availability of plastics waste management approaches in strategic management.
- One would be hard-pressed to find or identify environmental issues to investigate or contribute to that are related to the most important plastics waste issue of recycling.

Plastics waste management represents a growing problem in countries worldwide due to increased production and consumption of virgin plastic materials and additives. An increase in the worldwide consumption of plastics has led to the generation of plastics waste. This increase in the amount of plastics waste produced in large quantities proves the importance of its management. There is no definitive methodology for evaluating the environmental effects of plastics waste management systems, although the methods used for environmental assessment are broadly applicable. Piecing together the evidence on plastics waste management in the foregoing environmental problems, the verdict is clearly one of fairly consistent failure to maintain the level of the plastics production base. Therefore, the conclusion is usually some sort of future analysis of plastics—where they come from and where they go—which only ends by recycling or landfilling. This would actually help to understand how they differ from other solid wastes.

References

1. Grossmann, D., Hudson, J.F., Marks, D.H., Waste generation models for solid waste collection. *J. Environ. Eng. Div.*, 100, 1219, 1974.
2. Wertz, K.L., Economic factors influencing households production of refuse. *J. Environ. Econ. Manag.*, 2, 263, 1976.
3. *Initiating and Sustaining Water Sector Reforms: A Synthesis*, p. 38, World Bank, 1999, Washington, DC, USA, 1999.
4. Cardinali, R., Waste management: A missing element in strategic planning. *Work Study*, 50, 5, 197–201, 2001.
5. Seadon, J.K., Integrated waste management - Looking beyond the solid waste horizon. *Waste Manag.*, 26, 1327, 2006.
6. den Boer, J., den Boer, E., Jager, J., LCA-IWM: A decision support tool for sustainability assessment of waste management systems. *Waste Manag.*, 27, 1032, 2007.
7. Seadon, J.K., Sustainable waste management systems. *J. Clean. Prod.*, 18, 1639, 2010.
8. Demirbas, A., Waste management, waste resource facilities and waste conversion processes. *Energ. Convers. Manag.*, 52, 1280, 2011.
9. Benabboun, A., Brahmi, Y., Tilmatine, A., Bendaoud, A., Miloudi, M., Dascalescu, L., *Experimental investigation of a new tribo-electrostatic separation process for mixed fine granular plastics IEEE Industry Applications Society Annual Meeting, IAS 2013*, Lake Buena Vista, 2013.
10. Kress, C. and Franchetti, M., *Post-consumer plastic sortation with the use of electromagnetic separation methods for recycling*, ASME International Mechanical Engineering Congress and Exposition, San Diego, 2013.
11. Yang, X.H. and Lu, C.Y., Kinematic properties and test of plastics' optimal separation equipment. *Chin. J. Mech. Eng.*, 43, 2, 132–135, 2007.
12. La, M.F., Moroni, M., Cherubini, L., Lupo, E., Cenedese, A., Separation of plastic waste via the hydraulic separator multidune under different geometric configurations. *Waste Manag.*, 32, 7, 1306–1315, 2012.
13. Malinauskaite, J., Jouhara, H., Czajczyńska, D. *et al.*, Municipal solid waste management and waste-to-energy in the context of a circular economy and energy recycling in Europe. *Energy*, 141, 2013–2044, 2017.
14. Richmond, B., The Dynamic thinking: A behavioural context. *The Syst. Thinker*, 8, 6, 6–7, 1997a.

15. Price, J. and Joseph, J., Demand management - A basis for waste policy: A critical review of the applicability of the waste hierarchy in terms of achieving sustainable waste managemen. *Sust. Dev.*, 8, 2, 96, 2000.
16. Ross, D.E., An autobiography of solid waste management. *Waste Manag. Res.*, 36, 5, 405–407, 2018.
17. Guerrero, L.A., Maas, G., Hogland, W., Solid waste management challenges for cities in developing countries. *Waste Manag.*, 33, 220–232, 2013.
18. Blanchard, O. and Viñals, J., Chapter 1, Global prospects and policies, World economic outlook (International Monetary Fund), pp. 1–42, Washington, D.C. 20431, U.S.A. 2010.
19. Haque, A., Mujtaba, I., Bell, J.N.B., A simple model for complex waste recycling scenarios in developing economies. *Waste Manag.*, 20, 625–631, 2000.
20. Kanat, G., Demir, A., Ozkaya, B., Bilgili, M.S., Addressing the operational problems in a composting and recycling plant. *Waste Manag.*, 26, 1384–1391, 2006.
21. Merkl-Davies, D.M. and Brennan, N.M., A conceptual framework of impression management: New insights from psychology, sociology and critical perspectives. *Account. Bus. Res.*, 41, 5, 415–437, 2011.
22. Bebbington, J., Larrinaga, C., Moneva, J.M., Corporate social reporting and reputation risk management. *Account. Auditing Account. J.*, 21, 3, 337–361, 2008.
23. Marquis, C. and Qian, C., Corporate Social Responsibility Reporting in China: Symbol or Substance. *Organ. Sci.*, 25, 1, 127–148, 2014.
24. Michelon, G., Pilonato, S., Ricceri, F., CSR reporting practices and the quality of disclosure: An empirical analysis. *Crit. Perspect. Account.* 33, 59–78, 2015.
25. Luna Sotorrío, L. and Fernández Sánchez, J.L., Corporate social reporting for different audiences: The case of multinational corporations in Spain. *Corp. Soc. Resp. Env. Manag.*, 17, 5, 272–283, 2010.
26. Villanueva, A.B., The Politics of Waste Management. *J. Soc. Political, Econ. Stud.*, 205–220, Summer 1996.
27. Tchobanoglous, G., Theisen, H., Samuel, A.V., *in Integrated Solid Waste Management — Engineering Principles and Management Issues*, McGraw Hill, New York, NY, USA, 1993.
28. Kumar, S., Bhattacharyya, J.K., Vaidya, A.N., Chakrabarti, T., Devotta, S., Akolkar, A.B., Assessment of the status of municipal solid waste

management in metro cities, state capitals, class I cities, and class II towns in India: An insight. *Waste Manag.*, 29, 883, 2009.
29. Richmond, B., Operational thinking. *The Syst. Thinker,* 9(2), 6–7, 1998.
30. Patel, M.K., Jochem, E., Radgen, P., Worrell, E., Plastics streams in Germany—An analysis of production, consumption and waste generation. *Resour. Conserv. Recy.*, 24, 191–215, 1998.
31. Ayres, R.U., Industrial Metabolism, in: *Technology and Environment*, J. Ausubel and H.E. Sladovich (Eds.), National Academy Press, Washington, DC, 1989.
32. Erkman, S., Recent history of industrial ecology, in: *A Handbook of Industrial Ecology*, R.U. Ayres and L.W. Ayres (Eds.), Edward Elgar, Cheltenham, U.K., Lyme, MA, 2002.
33. Checkland, P., in: *Systems Thinking, Systems Practice*, Wiley, Chichester, UK, 1981.
34. Christensen, C.M. and Bower, J.L., Customer power, strategic investment, and the failure of leading firm. *Strateg. Manage. J.*, 17, 197, 1996.

7

Recycling Technology

Global production of polymers reached 322 million tons in 2015, 40% of which were destined for packaging applications [1]. The disposal of plastic waste has become a major worldwide environmental problem. Increasing amounts of waste plastics pose a threat to social development and public health [2]. Plastics waste are not biodegradable over time due to their mechanical strength and resistance to both water and microorganisms occupying a large surface of arable land [3, 4].

Waste plastics are usually disposed of in landfill sites. The landfilling of plastics waste is a short-term solution, which does not allow the recycling and recovery of any resource. The continued buildup of plastics waste leads to landfills. It has become a major source of pollutants for many sectors of society. It is up to the manufacturer to figure out how to diminish the problem and considerable technical skills are needed to segregate and utilize plastics waste as energy efficient material which can significantly contribute to economic development to improve quality of life. Modern society requires reasonable alternatives to effectively cope with needs for energy and environmental conservation. Efficiency and development results in an increase of both consumption of products and wastes. Therefore, recycling technology practices include

- Collection; this is the most important process and requires careful planning to reduce operating costs, due to its high costs and population growth,
- Processing,
- Disposal,
- Reduction of waste generation to save energy and cost.

Muralisrinivasan Natamai Subramanian (ed.) Plastics Waste Management, 2nd Edition (115–162) © 2019 Scrivener Publishing LLC

The use of plastics and the subsequent generation of plastics waste are both rapidly increasing worldwide. The control of plastics waste remains an almost unconquered challenges for plastics waste management. From the perspectives of both fundamental and applied polymer technology, the ability to carry out recycling at a high level is highly desirable [5]. Methods allowing recyling process to be carried out on ultra-small scales in a controllable fashion are potentially important for a number of reasons, including population growth. Ultra-small scale recycling also circumvent potential potential problems associated with plastics waste, energy consumption, and waste generation [6].

7.1 Man-Made Material - Plastics

As advanced technologies are employed to research and develop man-made plastics, it is necessary to invest considerable effort, energy and cost; however, these developed plastics should be strong, stable and durable. Plastics have become important materials for scientific and technological development, and contribute to maintaining a high standard of living; modern society depends upon polymeric products in everyday life, from toothbrushes to computers.

Plastics are designed to exhibit durability and strength; however, an unfortunate consequence is that they create environmental pollution after use, i.e., plastics do not easily degrade in the outdoor environment and accumulate at dump sites causing litter. The required properties for the end use of plastics results in their relative resistance to environmental degradation, which includes biodegradation. Plastics waste has been recognized as one of the most complex and cost-intensive services and is a main area of interest. The use of plastics and the subsequent generation of plastics waste are both increasing rapidly worldwide. Advances in polymers are frequently breathtaking. The concept of plastics recycling shows great promise and open the door for reduction generated prior to recycling. Although living standard can be generalized towards population [7] and economic growth [8], a generalized waste collection and recyling depends on different phase behaviour.

7.2 Substantial Prerequisite

The increase in global environmental loading causes landfill-related issues. Plastics waste is discarded after a relatively short period of use. The continued buildup of plastics waste leads to landfills. It has become a major source of pollutants for many sectors of society. Landfilling of plastics waste requires a lot of space, wiping away valuable land resources.

Plastics waste recycling has become an important policy. It reduces disposal by landfill or incineration by diverting end-of-life material to alternative productive uses. Landfill reduction was historically important due to perceived landfill capacity shortages remains important today and in future due to significant non-degradable nature of plastics waste. Material recycling is thought to divert end-of-life waste away from both disposal options [9].

7.3 Philosophy

The philosophy of recycling technology not only applies to large and technical operations, but also prevents the generation of waste. The sustainable use of plastics from resources is related to durability and the use of environmentally compatible materials. However, fresh raw material production generates in the form of waste until a solution arises. Even after application of the recycling technology principles of avoid, reduce, recycle and recovery, the fresh raw material generates further residual waste and becomes a permanent environmental burden in the case of plastics wastes. The quality and usage of plastic products, reuse, recycling and the technology of waste affect the performance of energy efficiency.

Recycling of waste plastics could be used to reduce landfilling, thereby reducing non-renewable resources usage.

- Recycling, reuse are key to recovery of plastics waste with high performance.
- Plastics waste can be a source of sustainable new material.
- Recycling methods to produce granulates is a way to obtain very cheap and useful materials.
- Plastics waste can be used to produce products or composites.

7.4 Purpose of Recycling Technology

The main purpose of recycle technology is to save resources: recycling reduces the need for virgin raw material, however, the contamination of plastics within the recycling stream by foreign materials is a major obstacle to recycling efforts. Many of the plastics processing industries have setup services, mainly targeting plastics waste to recover and recycle from the municipal solid waste.

Recycling technology is typically utilized and other measures such as reuse, recovery, etc., are used in the case of mixed or contaminated wastes. They are first considered for reuse and recycling, and the rest for recovery, and finally the residues are sent to the landfill [10].

Therefore, the purpose of recycle technology is to

- Minimize the amount of waste materials.
- Prevention and reduction of waste generation
- Maximize the recovery
- Greater protection of the environment and human health.
- Reduce operational problems
- Reduce raw materials consumption

Recycle technology is necessary to conserve and dispose of plastics waste as there is no alternative solution, hence it should be considered as a resource; disposal techniques depend on the type of waste and location. It provides an ecologically acceptable way of reutilizing the energy content of the waste. The plastics processing operation is damaging to the environment and inefficient if the polymer composition of the waste is unknown. It has been suggested that recycle technology is the only sustainable solution to the rapidly increasing amount of plastics waste.

An effective recycle technology not only reduces the environmental impact, but sends a positive message to its end users. The plastics waste produced in a manufacturing facility is the result of its processing operations and the easiest way to reduce waste is, most likely, to alter the process. The hierarchy of reduce, reuse and recycle is also applied to process waste. Making process changes in order to reduce waste, usually requires an investigation of costs, alternatives and results. Therefore, recycling technology is used to minimize the environmental burden from wastes and to save resources.

The raw materials from the recycled plastics are cheaper in comparison with virgin materials due to energy savings. Recycling technology encourages the recovery of plastics waste. It prevents or reduces waste by means of clean technology development. The product production contributes in the smallest possible way to recycling in order to reduce waste. Production of plastics comes to a halt once all recycled materials are consumed, thereby helping to preserve petroleum resources.

7.5 Fortune of Plastics Material

Recycling plastics waste is a fundamental requirement in decreasing the quantity of resources required from primary sources, as recycling feeds back into the supply chain.

- Lifetime of a plastics product is significant in evaluation of the supply risk. According to the length of time that a product lives, fewer product recycles are required in the given amount of time. This results in reduction of associated losses [11, 12].
- Mechanical strength of plastics material is related to the amount of virgin or recycled resources required for a specific application [13].
- Substitutability is related to the possibility of replacing virgin material in product with recycled plastics with equivalent function.
- End-of-life and recycling refer to advances in recycling technology which decrease the amount of waste. Technological developments at the product level refer to changes in the product design and increase in the module efficiency.

7.6 Methods of Recycling

Effective technologies and practices have stimulated industry, society, and authorities to develop the latest advancements in solid waste related to plastics waste. The ultimate goal of recycling technology is to reach maximum recovery while limiting transportation costs. Due to environmental issues, waste plastics recycling methods are under development in an economic and ecological way. Numerous processes have been developed to recycle plastics waste, two of which are

- Recovery of energy by incineration
- Combined recovery of energetic and chemical value

The knowledge integration in problem-solving solutions related to the environment would not be possible to solve within a single field of knowledge. Different areas, such as compounding, blending, processing, etc., illuminate the societal impact of waste plastics problems. Recycling technology translates problems into commercial success. Due to environmental issues, waste plastics recycling methods are done in an economic and ecological way. There are different approaches to plastics recycling, the main processes being [14–17]:

- Reprocessing by different techniques such as injection molding, extrusion, etc. These processes regrind scrap from industrial or single polymer plastics edges for use in processing to produce products of similar material.
- Mechanical recycling, in which material is recovered for reuse in manufacturing. This can be performed on individual plastics. No contamination is present during the recycling. Contamination or mix is difficult to recycle by mechanical means.
- Monomer separation of plastic materials into their constituent monomer or their petrochemical. The alteration in the chemical structure is bound to occur.
- Burning of plastics to produce energy in the form of heat, steam and electricity. Plastic materials possess a very high calorific value (when burned); especially when considering that they are derived from crude oil. For instance, energy recovery occurs either by incineration or combined recovery of energetic and chemical value.

7.7 Plastics Waste - Stream

Plastics waste originates from packaging, municipal solid waste, construction and demolition, end-of-life vehicles, medical and biomedical sources, and electrical waste and electronic equipment. The main difficulties encountered in recycling postconsumer plastics are: 1) most postconsumer wastes are soiled and 2) they exist in a dilute form in terms of quantity in the community [18]. Although the technology for washing and separating the various types of plastic is available, the number of facilities with this capability will continue to be extremely small until a sufficient supply of raw material, at an appropriate cost, can be assured. One remedy would be to develop material collection processes that are resource efficient.

The recycling of mixed plastics waste has attracted considerable interest because of the economic and ecological advantages compared with separately compounding each component, i.e., virgin polymeric materials and their additives [19, 20]. In general, deterioration of material properties, caused by the incompatibility of the components, is one of the major problems in reprocessing mixed plastics waste. Almost 85%, by volume, of global plastics consumption is comprised of four thermoplastic resins, PE, PP, PVC and PS [21].

The first step starts with determining the types and quantities of discarded materials. The selection of plastics recycling streams involves:

- Observation of the types and amounts of waste produced.
- Identification of waste-producing activities and equipment.
- Detection of inefficiencies in operations or the way waste moves through the organization.
- Assessment of space and equipment that can be used for storage, processing recyclables and other activities.
- Assessment of current waste reduction efforts.

Consumers of plastics generate waste in diverse fields such as domestic, hospital, industrial, etc. The two main drivers of the environmental interest in plastics waste are

- Control over the production of plastics raw material.
- Disposal methods to preserve the environment and exploit waste plastics.

Plastics contribute to continuous energy savings. They are inert and non-biodegradable and help in sustainable development. Economic barriers could be overcome once the products generated from plastics are more valuable and if mixed plastics could be processed together, eliminating sorting costs. Renewed waste plastics are valuable sources for manufacturing products with similar status to virgin materials [22]. Plastic waste contains certain amounts of polyvinylchloride (PVC) in products such as foil, bags, sheets, etc. Bottles, cups and cans are made from from PVC, high and low density polyethylene, or polystyrene.

Figure 7.1 illustrates the industrial recycling of discarded waste into raw material. In municipal waste, postconsumer plastic represents a small percentage; however, it occupies about 25% of the waste stream volume.

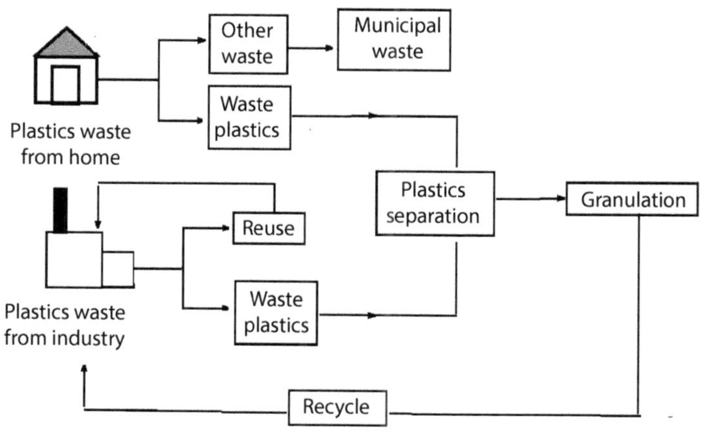

Figure 7.1 Plastics waste stream.

Plastics, with their increasing stream of waste, are subjected to recycling. The most common form of waste disposal is combustion. However, due to the generation of harmful gases, pyrolysis or gasification is used as an alternative for disposing of waste plastics [23].

7.8 Mixed Plastics Waste – Separation

In mixed waste plastics, the difficulties are technology, dismantling, identification and separation. To avoid dismantling equipment, such as a die during processing, requires scientific knowledge in the identification and separation of individual plastics in order to influence recycling. Plastics without any control cause pollution problems.

To convert waste plastics into powder form they are electrically charged by friction in the tribocharger, which has different types of spiral tubes, fluidized beds, rotational cylinders, etc. The electric characteristics of the charged particles depend on the chemical components. In powdered waste plastics, plastics separation can be possible based on the differential charges [24]. Figure 7.2 illustrates the separation of plastics from the mixed waste collection.

In mixed plastics waste, the combination of interesting material properties and environmental acceptability makes recycled products suitable for different consumer applications. Plastics waste recovery, with the exception of PET, from the waste stream is to be made compulsory. Plastics waste is subjected to a series of sensors that sort the type and color; further inspection ensures no PVC contamination.

Any waste which is contaminated by soil, or contains dirty plastics, cardboard and so on, cannot be recycled. Small quantities of low density materials such as low-density PE and PP can be mixed with high-density polyethylene (HDPE); a negligible quantity of PS is not an issue [25].

Mixed plastics separation by electrostatic separation method uses different electric properties of the materials. The method utilizes electrostatic force to separate mixed waste material characteristics.

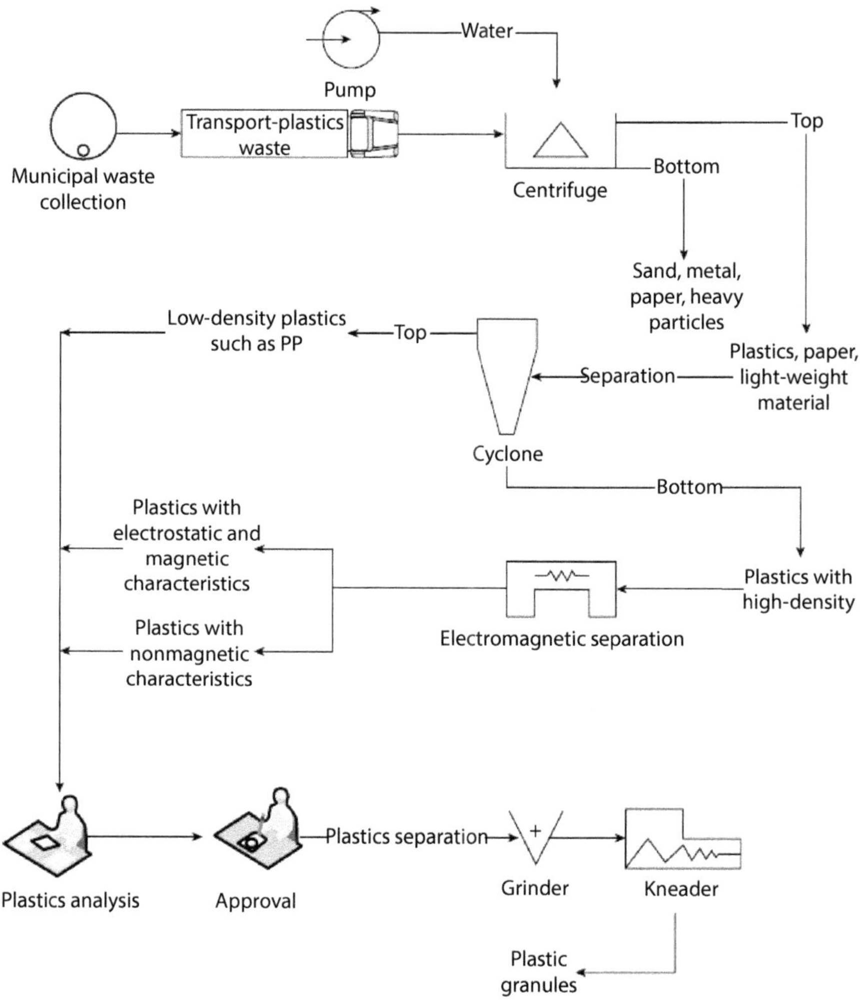

Figure 7.2 Plastics waste separation.

7.9 Origination of Plastics Waste

Plastics waste originates from

- The poor quality of plastic parts from solid urban wastes. The cost to purify places considerable constraints on the economic viability of recycling [26, 27].

- Incompatibility of different plastic materials present in the waste stream together with contamination by non-polymeric materials. Mixed or contaminated waste plastics results in poor mechanical properties. With too much contamination, sometimes the products are more inferior, which may be expected.
- The recalcitrant nature of plastic combined with a single-use throw away culture and release into the environment, results in negative environmental consequences [28].

7.10 Problems of Recycling and Controls

Plastics waste disposal poses a research challenge and on a global scale, plastic products are considered to be non-degradable. The majority of plastics waste is not produced from petroleum waste, but from natural gas that could be used for a different purpose or not extracted at all [29].

Plastics waste issues are becoming highly uncertain, urgent, and complex due to the interconnected supply of plastics raw material for processing. We can no longer afford to address individual environmental and social problems with respect to manufacturing of plastic raw material and their additives for processing.

7.10.1 Problems

- The set of process variables involve the validation of reprocessing issues.
- Plastics waste should not be contaminated, particularly with indications of other plastics or contaminants.
- Pessimistic utilization of plastics waste is a risk factor for further waste generation, which is a critical problem requiring immediate utilization of the materials.
- Plastics waste is discarded as part of municipal solid waste, construction demolition waste or single-use plastic products, the thin but visible litter stream.

- Existing financial resources are not sufficient for the collection, transportation, appropriate separation and disposal of plastics waste.

7.10.2 Controls

- Plastics waste should be assessed related to the suitability of recycling. Sorting and separation of individual plastic material are, clearly, useful options, but they take time and motivation.
- Industrial metabolism and ecology have motivated to identify plastics waste that could be recovered, upgraded and reused as alternative raw materials for industry, thereby closing the cycle of materials. Recycling of plastics waste provides an ecologically acceptable way of reutilizing the energy content of the waste.
- Recycle techniques for setting production and evaluation.
- Engineering and other technologies to recycle for plastics waste handling follows economic reality and higher scrap values are resulting in new market potentials. The energy emergence in environmental concern has highlighted the energy value of waste plastics as fuel.
- Plastics waste offers advantage of pre-sorting and grading from other solid wastes next to metals.

7.11 Physical Characterization and Identification

The processing of plastics waste requires methods of disposal which track the flow of material as it is essential to quantitatively evaluate if the recycling method employed is feasible or not. There is even a possibility that recycling may result in negative environmental loading. The recycled material has to be initially subjected to recognize testing methods as it is necessary to know the composition of plastics waste is essential in light of their diversity and heterogeneity.

There are three types of composition namely, physical, proximate, and ultimate analyses.

- Physical characterization involves identification of plastics.
- Proximate analysis involves the determination of percentage of carbon, volatile matter and ash.
- Ultimate analysis informs about the amount of the amount of carbon (C), hydrogen (H), nitrogen (N), sulphur (S) and oxygen (O).

Thermal characteristics such as heating value, in particular higher heating value are key factor for plastics waste to energy. However experimental determination is expensive.

7.12 Recycling – A Resource

During the plastic material life cycle, plastics waste is recycled as a resource. The amount and type of recyclable plastic depends upon the composition and quantity of waste from the municipal and industrial waste streams. Recycling is a process referred to as 'value from waste', which indicates the recovery or reuse of material via the extraction and reprocessing of raw materials from recycled products. Methods are constantly being developed which allow more effective ways of recycling waste materials. The continued buildup of plastics waste leads to landfills. It has become a major source of pollutants for many sectors of society. It is up to the manufacturer to figure out how to diminish the problem and considerable technical skills are needed to segregate and utilize recycled material. Recycling is associated with many complications, namely,

- Problems related to separation, sorting and cleaning operations,
- Lack of fiscal subsidies,
- Instability of selective garbage separation programs,
- High transport and electricity costs, etc.

7.13 Recycling Technology

Recycling technology has become more and more common for two different reasons [28, 30]; the first is driven by cost, which requires the reprocessing of faulty parts and scrap from molding methods. The second is environmental, which is crucial in the case of commodity plastics due to the large production volume. Recycling technology involves collecting, separating, processing, marketing and ultimately using material (Flowchart) that would have been discarded and is associated with community roadside programs.

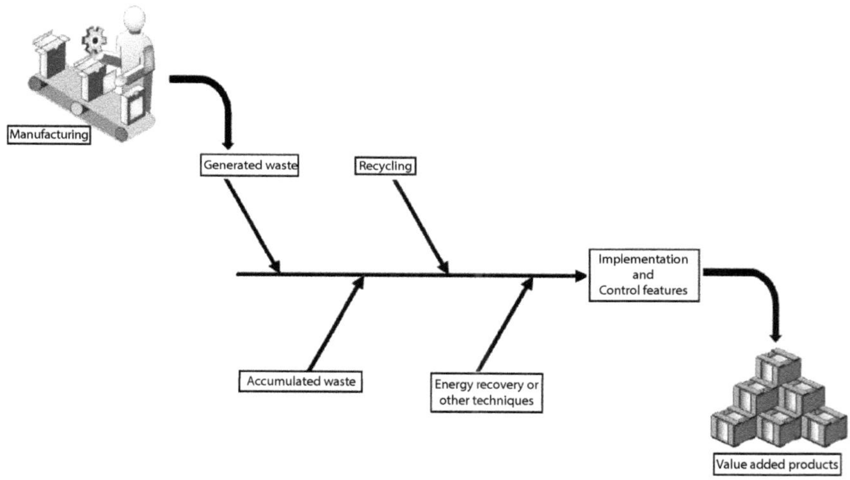

Flowchart – simple illustration of plastics production and recycle technology.

Either recycling or disposing of plastics waste puts inexorable pressure on the environment due to its very low biodegradability and presence in large quantities. The success or failure of an organization can be influenced by the type of innovations and developments which are available in the field of recycling plastics waste. Innovative recycling methods enable the transformation of plastics waste into new products and additional applications. Most proposals for the recycling or disposal of plastics waste require human and financial resources. Recycling addresses the needs of small industries in remote or emerging markets by enabling the development and progression of this sector. If the recycling process does not

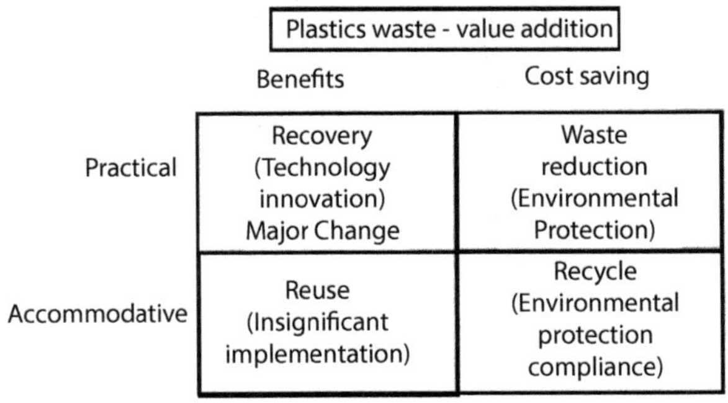

Figure 7.3 Practical and accommodative recycle oriented environmental policy.

move forward, these organizations will have insufficient recycled plastic and consequently lower productivity. There are four main approaches to plastics recycling; primary, secondary, tertiary and quaternary recycling [31–34].

Recycling technology provides industries with a framework for designing a practical and accommodative recycling oriented environmental policy (Figure 7.3). It enables industries to respond more adequately to their environmental performance.

The framework of an industry's environmental policy is tied up with its recycling policy. It is also directly related to the system being developed within the industry. By gearing recycling technology to the internal system of environmental technology, and vice versa, plastics waste is in principle less of a problem, as the gap between strategic technology and implementation of recycling is short.

7.14 Primary Recycling

In-house recycling is primary recycling which ultimately strives for primary closed-loop recycling of plastics, but has only been achieved for a very narrow plastic segment, such as clear PET bottles, for which dedicated collection schemes already exist. Industrial waste from the manufacture of plastic products is regarded as a manufacturing loss; the reduction of plastics waste, as a result of recycling the material, increases productivity and reduces the manufacturing loss. The in-house reuse of process waste has been common industrial practice for many years, however, it does not always divert waste from landfills. As disposal costs increase, the reuse of previously discarded waste from one industry by another is becoming more common. During processing, significant amounts of product end up in recycling bins and dumpsters, which is problematic.

Plastics waste recycling can help protect the environment and improve manufacturers' profitability. When plastics waste is recycled it is a valuable resource for manufacturing products, with a similar status to virgin materials. There is considerable profit in recycling plastics waste and the marketing of recycled plastics is not a problem due to the eagerness of end users to utilize the material.

7.14.1 Reprocessing Essentials

Before collection and transportation, separate storage of plastics from other types of waste is needed in reprocessing unit. However, it is to get a sufficiently clean and homogeneous plastic material. Reprocessing should have agglomerating equipment, granulators and extruders.

- Regeneration of plastic from plastics wastes
- Should be with satisfactory quality for production
- Should not have contaminations other than plastics which reduces the plant capacity and lower revenues.
- Require mechanical wet refining and/or sorting before regeneration to prevent equipment clogging from contaminations.
- To avoid with unnecessary refining which reduces the volume of the recycle plastic.

7.15 Mechanical Recycling

Mechanical recycling is a secondary recycling of plastics waste into products of inferior qualities, but is limited mainly to fractions containing only a single type of plastic. The lack of primary recycling of plastics is challenged by many aspects, such as colors (dyes) and other contaminants, qualities, blends of different plastics, etc., and is manifested by the fact that all plastics are closed-loop recycled. Mechanical recycling (or material recycling), is the process of recovering plastics for the re-use in manufacturing plastic products via mechanical means [32–36]. Mechanical recycling of plastics can only be performed on single-polymer plastic. The more complex and contaminated the waste is, the more difficult it is to recycle it mechanically.

Mechanical recycling is also known as re-extrusion, is the re-introduction of scrap, industrial or single-polymer plastic edges and parts to the extrusion cycle in order to produce products of the similar material [33–34]. Mechanical and feedstock recycling involve material recovery from plastic waste streams and combustion procedures for heat production with controlled emissions, respectively, which are included in recycling technology as energy recovery options. The mechanical recycling of waste entails a physical means of producing new plastic products. Post-consumer plastic waste as virgin or second grade plastic feedstock is achieved using advanced mechanical recycling. Recycled plastics can be transformed into pellets which can be used as raw materials to either partially or fully replace virgin material in many plastic manufacturing operations.

Mechanical recycling is performed by physical means and involves separation, washing, grinding, remelting and processing plastics waste, and is a clean and homogeneous route for waste plastics. LCA studies indicate that mechanical recycling is preferable to other technology procedures in terms of optimizing overall energy use and minimizing the emission of gases that contribute to global warming [37].

Successful mechanical recycling of waste plastic is based on

- The properties of the recycled material depend upon the history of the plastics, the primary processing and application, and finally the recovery methods/equipment.
- The purity and uniformity of the plastics waste.
- The composition of plastics waste is closely monitored prior to recycling as many plastics are not recyclable due to the presence of mixed plastics waste.
- The recycling of plastics is only economically viable when the processing system is highly efficient and cost-effective [38].

- Sorted, single polymer waste exhibiting low contamination stands a good chance of being recycled. If small amounts of other polymers, such as PE produced using a Ziegler or Philips-type catalyst, PP [39], even lower amounts of polystyrene (PS), polyethylene terephthalate (PET) and polyvinyl chloride (PVC) are present, the waste cannot undergo commercial recycling [40].

7.15.1 Limitations

Recycled products from mechanical recycling are exposed to the mechanisms and effects of degradative processes throughout their life cycle. Using analytical strategies, the quality assessment of recycled plastic includes determining the degree of mixing (composition) and degradation, such as chemical and morphological alterations, changes in the mechanical and rheological properties, and the presence of low molecular weight (MW) compounds, such as contaminants, additives and degradation products [22, 41].

- Mass recycling of materials is not suitable due to a loss of material properties.
- The contamination of plastics within the recycling stream, by foreign materials, is a major obstacle to recycling efforts.

Waste plastics with improved properties can be realized by the restabilization and rebuilding of the molecular structure and compatibilization of mixed waste plastic blends. Effective mechanical recycling takes into account specific technology- and technology-related issues. Upgrading the plastics waste process is a necessity both for environmental protection and sustainable development. The incineration of waste plastic is declining due to operating costs and descending credibility in the environmental protection scenario.

7.15.2 Processing Problems

The recycling of plastics waste in mechanical recycling is often limited by the degradation which occurs during processing. The effects of degradation include: reduced physical properties, surface defects, and process instability resulting in machine downtime and wear, and increased quality control costs. Several types of degradation are known to occur during processing, namely: thermal, mechanical, chemical and oxidative.

Thermal degradation in polymers occurs at elevated temperatures. Mechanical degradation is due to excessive shear during processing. Chemical degradation is caused by a reaction between the polymer and a chemical. Oxidative degradation arises as a result of heat, oxygen and mechanical parameters [42]. Due to adverse weather conditions and transportation problems, recycled plastics may quickly become depleted due to the rate of waste collection and sorting of the municipal waste streams.

7.16 Chemical Recycling

Since industrial revolution, the world has been heavily dependent on petroleum and natural gas resources. Chemical recycling is a tertiary process may provide a solution towards the energy crisis. Chemical recycling conforms to the principles of sustainable development. It leads to the formation raw materials from which the plastics are originally made. It is able to recover the energy content of the plastic in the form liquid and gas.

Chemical recycling, involving decomposition or depolymerization of the polymers for recovery of monomeric constituents or other valuable chemicals, is a prospective method for fractions of plastics where primary and secondary recycling is unviable [32]. Chemical recycling includes chemical depolymerization, solvolysis, catalytic and thermal cracking, pyrolysis, gasification, hydrogenation, etc. [43]. The future vision of the economy and plastics waste management will stimulate an evolution in the approach to recover/generate not only energy but also monomer recovery.

Chemical recycling returns valuable products from waste. Thermal cracking of thermoplastic waste is well-known technique. Chemical recycling involves cracking and depolymerization by chemical process. This forms a series of petrochemical products or monomers. The products from the feedstock process can be transformed into new polymeric products by synthesis. Cracking of LDPE dissolved in benzene solution is shown in Figure 7.4.

Chemical recycling is the transformation of waste plastics into small molecules. It involves cracking and depolymerization by chemical process. The term chemical is used due to the fact that an alteration is bound to occur in the chemical structure of the polymer [44]. Chemical recycling converts plastic materials into their constituent monomers (monomer recycling) or to their petrochemicals (called feedstock recycling), usually liquids or gases, which are suitable for use as a feedstock for the production of new petrochemicals and plastics. This forms a series of petrochemical products or monomers. The products from the chemical recycling process can be transformed into new polymeric products by synthesis. They are suitable for use as products for the production of new petrochemicals and polymers.

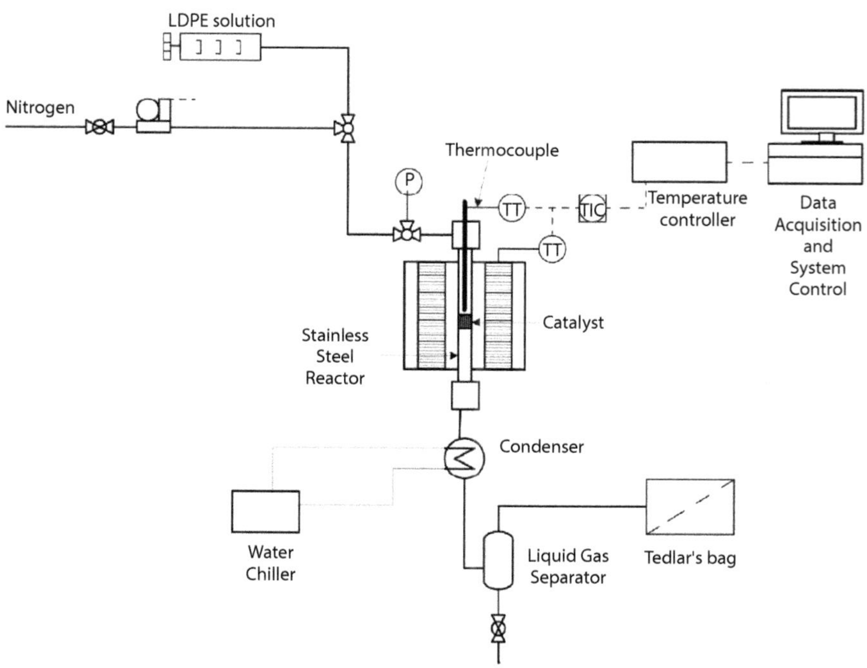

Figure 7.4 Schematic diagram of the LDPE cracking system [Reprinted with permission from S. Wong, N. Ngadi, T.A.T. Abdullah, and I.M. Inuwa, Catalytic cracking of LDPE dissolved in benzene using nickel-impregnated zeolites. *Ind Eng Chem Res* 2016;55:2543–55. Copyright © 2016 All rights reserved].

During pyrolysis, waste plastics undergo chemical change involving cracking phenomena. There are three major types of cracking namely random, chain strip and chain cracking [45]. Depending on the process like slow and flash pyrolysis, a number of reactor systems have been developed. Pyrolysis of plastic wastes, it is possible to recover hydrocarbons like olefins, aromatic and aliphatic compounds [46–49]. Indirectly heated fluidized bed gives a smaller number of different compounds due to its good heat and mass transfer properties. However, the yield of olefins produced from polyolefin does not exceed a maximum of 60 per cents.

Polymers like polyesters, polyethers, polycarbonates, and polyamides, are prone to solvolysis, such as hydrolysis (thermal and/or catalytic); others are more chemically resistant and require severe thermal conditions in order to decompose. For such polymers, including polyethylene and polypropylene, pyrolysis is suitable for converting the resins into basic chemicals and oils [50, 51].

The polyethylene can be thermally cracked or depolymerized back to lower molecular weight polymers or even, under severe conditions, to oily materials or to carbon and hydrogen. This process has been proposed as a method of recycling. Monomer recovery with specialty plastics has been remarkably technically successful on a large scale.

In case of PMMA and PS as feedstock, the main components are the monomers [52]. The fluidized bed pyrolysis of PMMA gives very large amount of monomeric methyl methacrylate under optimal conditions. Short residence time of PMMA in the laboratory experiments provide higher yield, however, it is hard to realize in technical fluidized beds. These polymerizations are inhibited by even small amounts of methyl butyrate.

Monomer recovery from PMMA is achieved by fluidized bed process. It has been a remarkable technically successful process for chemical recycling of plastics waste. The relatively high cost of PMMA compared to other standard plastics is meant for recycling with higher recovery of about 97 percent possible by fluidized bed process. Styrene monomer is produced in addition to recovery of percent oligomer formed during the depolymerization process of polystyrene.

7.17 Energy Recovery

Energy recovery is a quaternary recycling where all in-house processing, mechanical, and chemical recycling not possible to perform. Plastics mainly constitutes hydrocarbons. They become a good source of energy. There is a possibility of energy recovery during disposal. There are many techniques available for plastics waste disposal such as controlled incineration, burning in kilns, and land filling. However, these ways are having their own inherent demerits like environmental pollution, inefficient thermal recovery, and wastage of useful land.

Energy recovery is also known as thermal recycling. The combustion of waste plastics produces energy in the form of heat, steam and electricity. Plastics have a high calorific value in comparison with other materials. The majority of plastic materials are mostly essentially made from crude oil. Deriving fuel from plastic waste in an efficient manner can reduce the fossil fuel depletion. Using derived fuel as alternate fuel in internal combustion engines. Energy recovery by pyrolytic transformers basically comprises of thermal-chemical decomposition at elevated temperature in the absence of oxygen.

Energy recovery implies burning plastics waste to produce energy in the form of heat, steam and electricity. Plastic materials possess a very high calorific value (when burned); especially when considering that they are derived from crude oil. Thermal treatment of recycling of plastics on a pilot plant scale yields a fuel gas and liquid products. This fuel gas is mainly used to heat an auto-thermal reactor. The liquid products obtained from the plastics waste during direct pyrolysis replaces partially the coal by 5–10 percent.

Energy recovery options utilize plastic waste streams as fuel for energy production, exploiting the high calorific content of plastics. The life cycle assessment (LCA) governs how the successful recycling of waste plastics is managed, i.e., recycle technology is driven by the technological process used to recycle waste plastics.

7.18 Pyrolysis

Pyrolysis is thermos-chemical decomposition of organic materials at high temperature in the absence of oxygen. The term is used instead of pyrolysis. Thermal pyrolysis refers to the decomposition of polymers due to the application of heat alone. Catalytic pyrolysis is the process where catalyst is utilized to alter the reaction mechanism of pyrolysis, hence the products yield and composition.

The energy approach is more attractive due to ready availability of power boilers capable of handling municipal refuse after relatively simple processing. Pyrolysis produces oil, gas or chemical intermediates. The most efficient use of raw materials as matter of survival is produced with seemingly unlimited resources. These biodegradable plastics from natural resources, such as lactic acid, glucose, etc., are seemingly limited resources [53]. In the absence of oxygen, plastics can be fragmented at high temperature to form oligomers due its long polymer chain.

Pyrolytic processes have been carried out by different types of equipment such fluidized beds, rotary kilns, and rotating reactors [54, 55].

- Recovery by incineration technique
- Combined energetic and chemical value by pyrolysis, thermolysis or gasification are names associated with the process

Some pyrolytic processes cannot operate when the plastic is mixed with other contaminants. Materials such as aluminum/polymer laminates, used mainly as packaging for food and beverage and other products such as toothpaste, are often found in MSW and do not currently have a suitable feedstock recycling method; most waste of this kind ends up as landfill.

Among the various plastics recycling methods, the thermal and/or catalytic pyrolysis of plastics waste into fuel show the greatest potential for a successful future commercial polymer recycling process, especially as plastics waste can be considered as a cheap source of raw materials in times of accelerated depletion of natural resources. The catalytic degradation of plastics waste offers considerable advantages [56].

Pyrolysis of plastics waste is thought to recover energy and fuel due to their high volatile content (92%) and calorific value (40 MJ/kg) [57]. Pyrolysis is conducted in the absence of oxygen to form three fractions: gas (mainly formed by light hydrocarbons -C1-C4-, hydrogen, CO and CO_2); a wax/oil fraction consisting of a mixture of aromatic and aliphatic compounds; and a solid residue [58–62]. The proportion of each fraction and exact composition varies depending on feedstock. For instance, gas contains CO and CO_2 if PET is present in the feedstock [63]. PS and PET yield an aromatic yield while HDPE, LDPE and PP wax is formed by aliphatic hydrocarbons in the C11-C20 range [64]. On top of that, temperature, heating rate, reaction time and the addition of catalysts also play an important role in product yields [65].

HDPE pyrolysis product yield and composition varied greatly with temperature, i.e., from 72.3 wt% wax and 22.1 wt% gas at 650 °C to 13.5 wt% wax and 76.16 wt% gas at 730 °C). HDPE thermal pyrolysis at 500 °C yields mainly wax in the form of C+20 aliphatic hydrocarbons (67 wt%) followed by C12-20 hydrocarbons (25 wt%), C5-C11 hydrocarbons (5 wt%) and around 3 wt% containing methane (0.03 wt%), C2 (0.15 wt%), C3 (0.57 wt%), and C4 (2.27 wt%) 21. The distribution of hydrocarbons in the wax fraction makes it a low quality product due to the high viscosity and high boiling point temperature range and so it is not suitable for use as fuel without further processing [66].

The high viscosity of the wax fraction prevents their flow downstream and may cause blockages along pipelines unless heated, thereby increasing operation costs. Since the gas fraction is also low in valuable products (hydrogen or ethylene for instance), the profitability of conventional HDPE pyrolysis is low.

Pyrolysis could be a solution for a better environment when applied towards plastics waste. It allows the recovery of both energy and organic content in cases where physical and/or mechanical recycling could not be implemented. Plastics have high calorific value and combustion produces only heat energy. However, energy recovery from plastics waste is useful as a fuel for energy production. Emissions are included in the energy conversion.

Pyrolysis, thermolysis or gasification are all associated with combined recovery of energetic and chemical value. They are also referred to as pyrolytic processes. These processes were conducted using homogeneous waste and pure plastics with different types of equipment, namely fluidized beds, rotary kilns, and rotating reactors. However, the processes were also conducted with composition limitation found in municipal waste or other waste with existing contaminations.

Pyrolysis under controlled condition is the key point of the combustion process to generate energy (Figure 7.5). In an incinerator, the plastics waste depends on heat energy and mass transfer of ash or residue.

The factor affecting pyrolysis is the waste bed height in the reactor. In cases where solid additives are present in plastics waste are reactive, the problem is more complex due to the interaction between the waste plastics decomposition, chemical reactions, and the temperature. In principle, there are different thermal processing methods possible for pyrolysis and gasification apart from conventional refuse incineration methods. Europe is developing recovery methods for municipal waste and the United States is focusing on the development of energy production such as the preparation of refuse-derived fuel from waste.

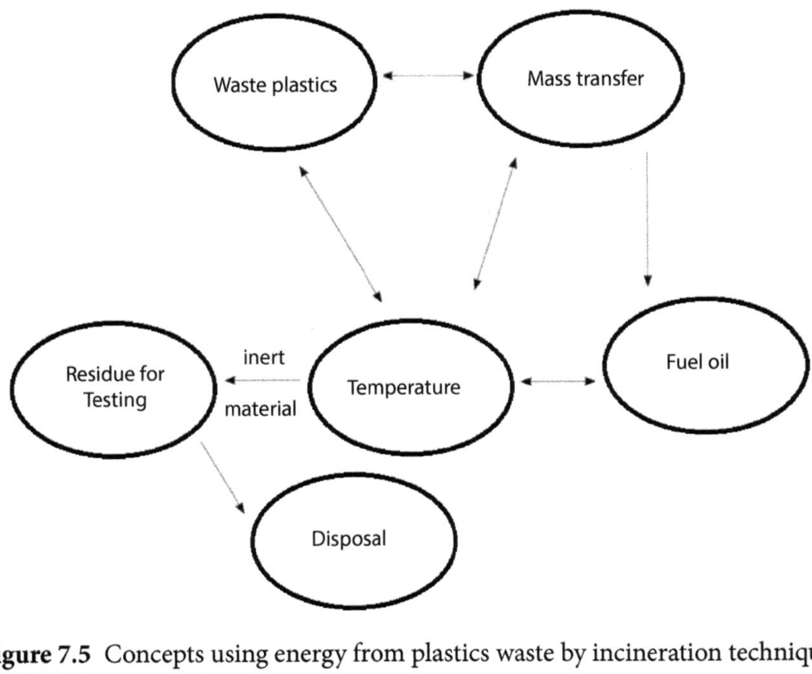

Figure 7.5 Concepts using energy from plastics waste by incineration technique for better energy recovery.

7.19 Types of Reactors and Process Design

Pyrolysis reactor design factors are based on [67, 68]

- Knowledge of the kinetics
- Thermal decomposition of the plastics waste

The mechanism of thermal degradation of plastics waste is very complex and includes other reactions such as chain fission, radical combination, carbon-hydrogen bond fission, hydrogen abstraction, mild-chain β-scission, radical addition, etc. [69, 70].

7.19.1 Batch and Semi-Batch Reactor

Batch or semi-batch reactors for thermal and catalytic pyrolysis of post-consumer plastic waste. It is easy to control the process parameters. The temperatures used for the process ranged from 300–900 °C and the reaction time from 30–90 min [71, 72]. It is possible to perform plastics waste at a temperature of around 200–275 °C. Some of the important parameters are

- Reaction temperature
- Mass ratio to plastic
- Catalyst
- Reaction time

With catalytic pyrolysis, mass ratio of reactant to catalyst varies and the ratio used is based on properties exhibited by selected catalysts and to increase subsequent productivity and yield [73, 74]. During pyrolytic process, the production of condensable vapor and fuel gases exerted pressure on the pressure valve. The valve opens at certain pressure to release these products into condenser and collection systems. The condensed oil is mixture of hydrocarbons in the range of kerosene and petrol and could be used as fuel after upgradation. On the reaction conditions, the exertion of pressure on the reaction system by the vapor and gases that increases the products yield at low temperatures.

Recycling activity in small scale pyrolysis will greatly increase the rate of plastic waste recycling, the sorting, cleaning and pyrolysis of the plastic waste can be carried out at location, hence transportation, storage, and pretreatment of waste prior to pyrolysis can be eliminated. Batch reactor are labor-intensive in its application in recycling industry [75]. However, with continuous pyrolysis provides the following advantages, such as

- Does not require frequently materials charging and restarting the process.
- Less labor-intensive

7.19.2 Fluidized Bed Reactor

Plastics waste in the reactor [Figure 7.6] at constant temperature is to possible to produce a narrower and more uniform spectrum of products by manipulating the residence time in the reactor [76]. It is possible to replace used catalyst periodically with process continuation

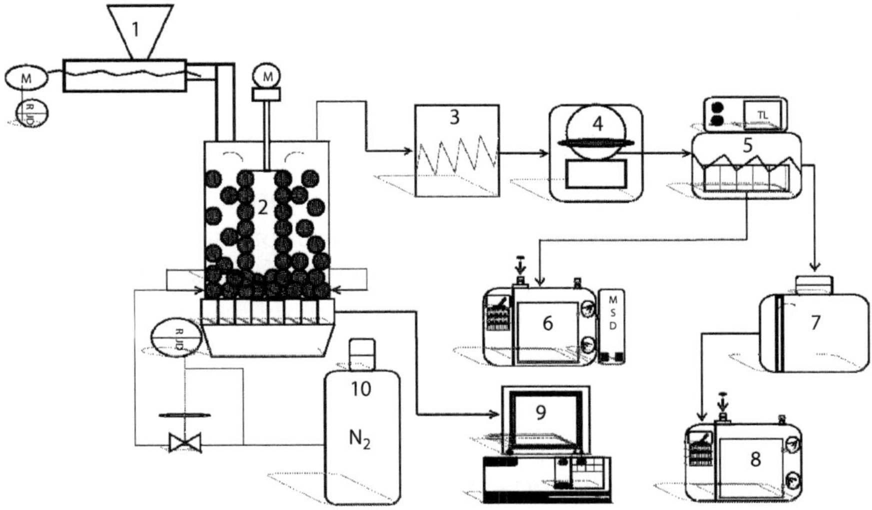

Figure 7.6 Schematic diagram of a catalytic fludizing system for LDPE/HDPE/PP. 1. Feeder, 2. Reactor with digital controller for three zone furnace, 3. Condenser, 4. Online multiloop automated sample system, 5. Separator, 6. GC/MS analysis of liquid products, 7. Gas bag, 8. GC analysis of volatile products, 9. TG analysis of residue products and 10. Nitrogen fludizing gas with mass flow control. [Figure reproduced with permission from [86]. Copyright © 2010 Elsevier, All rights reserved].

[77]. However, care has to be taken in order to avoid defluidization, as the process can easily when melted plastic stick on the bed.

Advantages

- Excellent mixing properties
- Improved heat transfer from reactor to polymer, compared to batch reactors [78].

Thermal pyrolysis of plastics on a fluidized bed reactor could produce higher degree of cracking compared to catalytic pyrolysis in a fixed bed reactor [79]. When considering the reduction of costs in catalyst replacement and regeneration, this process might be feasible in term of economy. Thermal process becomes less attractive in plastics waste recycling with the development of catalysts.

7.19.3 Conical Spouted Bed Reactor

To avoid defluidization caused by melted plastics in fluidized bed reactor, the use of this reactor provides vigorous contact between different phases and collection between particles reduced the chance of particle agglomeration in the bed materials. The extremely short residence time of plastic material in the reactor is also important to avoid secondary reactions which lead to formation of coke precursor [80]. The advantage of this reactor is lower pressure drop. When continuous thermal pyrolysis performs on plastic waste at moderate temperature range in CSBR [81]. The existence of secondary and tertiary reactions during thermal pyrolysis, producing more hydrocarbons with shorter chains.

7.19.4 Two-Stage Pyrolysis System

Despite of the advantages mentioned in the CSBR pyrolysis, the process tends to produce more wax rather than liquid and gaseous products. In such design, the plastic waste is melted and thermally cracked in the reactor at the bottom, and the vapor formed was then flown upwards and reacted with the catalyst on the fixed bed reactor. Such design avoided the direct contact of the catalysts with melted plastic, which causes several problems: difficulty in catalyst recovery after pyrolysis process, and rapid deactivation of catalyst when contacted with melted plastic, as well as impurities that are present. In such system, conversion of the polymer depended only on the temperature of the thermal reactor [82, 83].

7.19.5 Microwave-Assisted Pyrolysis (MAP)

Recently, microwave-assisted pyrolysis is receiving increasing attention from researchers, due to several advantages it has over the traditional pyrolysis, including faster heating rate and cost effectiveness. One distinctive feature in MAP is the application of microwave power on polymer materials, which is converted into heat at a fast rate. It is possible to raise the reaction temperature up to 1000°C in a short period of time in MAP using microwave absorber, usually carbon material or metal [84].

HDPE and toothpaste packaging, which was a combination of aluminum and PVC. It increases reaction temperature shortened the reaction time, but not have much effect on the products distribution. However, it is more important to know that MAP successfully apply to the Al/PVC mixture. PVC pyrolyze to form products with distribution similar to that of HDPE [85–87], while aluminum left in the reactor with almost 100% recovery, which could be separated easily from the carbon, the absorbent in the process.

The process reveals that higher microwave power causes more efficient degradation of polymer [88]. The possibility of MAP to convert commingled polymer waste that cannot be separated from the impurities using manual sorting, which is another advantage of MAP over traditional pyrolysis.

The shape of the absorbers plays important part in MAP, as it affects the amount of heat generated from the microwave. Microwave-assisted co-pyrolysis of waste PS with coal was also reported to produce aromatic liquid products in narrow range, together with acetylene and H_2S as gaseous products [89]. The performance of MAP on different types of plastic waste, as well as the effect of catalysts in MAP. The effect of interaction of different types of plastic waste on the products distribution in MAP is also an important feature to be explored.

7.19.6 Pyrolysis in Supercritical Water (SCW)

The use of sub-critical water (CW) system and supercritical water system (SCW), which are usually used in cracking of heavy oil [90]. In plastic recycling, as supercritical water behaves as solvent and catalyst simultaneously in the process [91, 92]. The small hydrocarbons molecules formed during polymer pyrolysis can easily disperse in SCW, hence the chance of subsequent condensation and coke formation is reduced [93, 94]. Excellent property of debromination was also discovered during pyrolysis of brominated polymer waste [95, 96]. The use of SCW in polymer waste pyrolysis also produce negligible amount of NOx, SOx and particulate emission [97]. Extension of reaction time results in improved liquid product quality, but coke formation also increases.

7.19.7 Fluid Catalytic Cracking

The development of pyrolysis technology, there is a need to develop continuous process for polymer pyrolysis before it can be applied in scalable production in the industries. However, one of the challenges faced is the continuous feeding of plastic waste in solid form into the reactor. In order to solve this problem, the possibility of dissolving plastic feed in suitable solvent, followed by pyrolysis of the solution [98].

The plastic dissolution in solvent is to incorporate the tertiary recycling process in the fluid catalytic cracking (FCC) process in petroleum refinery [99]. The advantages of using fluid catalytic cracking in polymer pyrolysis include [100]:

- Dissolution of plastic waste in the solvent increases its bulk density, hence reduce the volume of the plastic waste prior to the process.
- The contaminants (labeling, dyes etc.) can be removed during the dissolution stage, hence the negative effect by these contaminants towards pyrolysis performance can be reduced.

Separation of mixed plastic to different components is believed to be vital, as it is often reported that pyrolysis of single type of plastic is higher than that for mixed plastic. Value added activities for petroleum refining companies that engaged in this type of plastics recycling.

Limitations in large scale applications are

- Pretreatment is needed to clear the contaminants in the municipal plastic waste.
- To ensure the efficiency of the process, the presence of the contaminants has to be minimized during the solvent recovery at the expense of other useful products that can be extracted from the solution.
- Many of the solvents are volatile, dissolution process has to be done with extra safety measures.
- Selection of suitable solvent due to pyrolysis mechanism is affected and hence, solvent has to be selected to ensure positive influence towards the products yield and distribution.

7.20 Thermal Co-Processing

Mixed plastics on thermal treatment appears to be very promising recycling process. Pyrolysis on pilot plant scale yields fuel gas mainly used to heat an auto-thermal reactor and liquid products used to produce fuels or chemical feedstock. Coal co-coking with plastics waste in used in the form of blends. With ever-increasing oil prices, recycled plastics are becoming an economical alternative for the production of a wide range of commodity plastic parts. Polyolefin polymers, such as polyethylene (PE) and polypropylene (PP), contain approximately 14% hydrogen; these materials could provide the hydrogen required for thermal co-processing with biomass, which could lead to an increase of the liquid production of oligomers or short chain polymeric materials.

Co-pyrolytic techniques have received considerable attention in recent years as they provide an attractive way to dispose of, and convert, polyolefins (PO) and coal into higher value fuel; the specific benefits of this method potentially include the reduction of waste volume, the recovery of chemicals and substitution of fossil fuels [101–103]. Studies have shown that mixed plastic wastes can be used as a minor component in coal blends without any detriment to coke quality [104]. Recently, Nippon Steel Corporation (Chiyoda, Japan) began the industrial application of PVC-free plastic waste recycling at the Nagora and Kimutsu Works (Aichi, Japan).

7.20.1 Advantages

- Many plastic components are used as refuse-derived fuel or as a reducing agent for blast furnaces [22].
- Enhanced recyclability or energy recovery with reduced operational problems.
- Recovery of plastics from waste with environmentally and cost-effective technology such as granulation for storage and stock as better recycled material.
- Unusable waste plastics can be gasified using thermo-chemical processes.
- Technical reliability of feedstock such as waste plastics from packaging.

7.21 Gasification

Unusable waste plastics are gasified using different gasification-based treatments such as pyrolysis and gasification, gasification and combustion, and biochemical techniques to convert them into fuel and gas. Direct combustion of waste plastics could be controversial. Indirect combustion is a combination of gasification of waste plastics and combustion of the obtained fuel gas. Air injection enables good mixing of combustion air with combustible gases formed by gasification in the fuel bed in combustion chamber. Environmental advantages of burning a gas and not waste plastics are generally to reach visibility as conventional solution [105]. The reduced amount of waste plastics requires to be used and appear competitive in combustion could favour thermal treatments like gasification [106].

- Evaluation of the environmental sustainability of waste plastics in terms of energy recovery efficiency, reduction of carbon monoxide emission, consumption and environmental indicators as tool.

- Enhanced recyclability or energy recovery with reduced operational problems.
- Recovery of plastics from waste with environmentally and cost-effective management such as granulation for storage and stock as better recycled material.
- In case of unusable waste plastics into gasification using thermochemical processes.
- Technical reliability of feedstocks such as waste plastics from packaging.

7.22 Plastics Waste and Recycling

Plastics waste generation will develop in the medium or long term with uncertainty. It is quite probable that the consumption activities of plastics products in absolute terms will continue to grow rapidly [107]. Integrated systems involve a recycling technology which maximizes the recovery of resources from the waste. Single-use plastic products are spread as disposed plastics waste. Over the past decade, uncontrolled use of single-use plastic materials for syringes have been accumulating, leading to the loss of biodiversity and destroying habitats of living organisms. The widespread use of plastics represents a risk to the environment and human health, especially those classified as medical plastic wastes.

7.22.1 Polyolefin

A major source of plastics waste comes from the polyolefin family due to their large consumption as packaging material. Moreover, plastics are hydrophobic nature, hence possibility of cleaning of sand and other materials stick with it. However, the oil stick with these materials are difficult. The oil contains oleo-chemicals and possible to recycle without any difficult.

In plastics processing, it is well known that the effective recycling of plastics in general, and of HDPE in particular, requires information regarding the polymer's properties [108]. The properties, in comparison with their virgin counterparts, depend upon the plastic type and the number of cycles and conditions that occur during reprocessing. The polyethylene can be thermally cracked or depolymerized back to lower molecular weight polymers or even, under severe conditions, to oily materials or to carbon and hydrogen. This process has been proposed as a method of recycling. Polyethylene is the most common material used in packaging. It contributes to the largest portion of consumer plastic wastes. It is mainly used in shopping bags, shrink foil, refuse sacks, etc.

7.22.2 Polyvinyl Chloride

The mechanical and processing properties of recycled polyvinylchloride (rPVC), collected from bottles and pipes, blended with pipe-grade unplasticized polyvinyl chloride (uPVC) were found to be acceptable. Two important factors which control uPVC processing are the particle size and restabilization. The impact strength and processing behavior of uPVC improves upon the addition of rPVC due to the presence of modifier agents in the rPVC pipes, which slightly lowers the thermal resistance [109].

Polyvinylchloride (PVC) and coal have different main components. The calorific value of PVC is higher than that of coal. The volatile contents are higher in PVC with low carbon. However, the hydrogen content is higher, therefore, it produces only water and a low level of carbon dioxide during incineration. Hence, PVC can reduce the emissions of carbon and mitigate the greenhouse effect. The sulfur content is much lower than that of pulverized coal. The use of PVC instead of pulverized coal can reduce the emission of sulfur dioxide. PVC is a solid fuel that is clean, green and efficient [110].

7.22.3 Polyethylene Terephthalate

PET waste is recycled using various methods for use in several applications. Chemical recycling enables sustainable development and leads to recovery of the raw materials from which the polymer was made, as well as secondary value-added products [111]. Chemical recycling is generally based on breaking the ester bonds of waste PET using a chemical degradation process such as [112]:

1. Methanolysis: converting PET into dimethyl terephthalate and ethylene glycol (EG). The products of

methanolysis are the raw materials needed for the production of PET. The reaction conditions for methanolysis are 2–4 MPa and 180–280 °C [113–116].

2. Glycolysis: chemical recycling using EG, diethylene glycol, propylene glycol, butanediol and triethylene glycol [117, 118]. The temperature range of the glycolysis process is 180–250 °C [119]. Glycolysis results in the production of specialized products, such as unsaturated polyester, polyurethane and polymer concrete; glycolysis is primarily used for this type of recycling due to its low cost [120].

   Glycolysis consists of the transesterification of PET and the destruction of its polymer chain, resulting in the decrease of its MW. When glycols are used for the depolymerization of PET, the oligomers obtained have two hydroxyl end groups, i.e., oligoester diols are formed.

   Unsaturated polyester resins are complex polymers resulting from a crosslinking reaction of a liquid unsaturated polyester with vinyl-type monomers, most often a styrene monomer. The unsaturated polyester is formed via the condensation reaction of an unsaturated dibasic acid or anhydride, a saturated dibasic acid or anhydride, and a polyfunctional alcohol [121].
3. Hydrolysis: acids react with PET producing dicarboxylic acid and diols which yield raw materials that can be used in production.
4. Ammonolysis: anhydrous ammonia reacts with PET producing terephthalic acid amide. Reaction conditions include a temperature range of 120–180 °C, 2 MPa and 1–7 h duration [122].
5. Aminolysis.
6. Other processes.

Plastics waste disposal poses a research challenge and on a global scale, plastic products are considered to be non-degradable. The majority of plastics waste is not produced from petroleum waste, but from natural gas that could be used for a different purpose or not extracted at all [123].

7.23 Environmental Burdens

Environmental burdens such as air pollution, water pollution, etc., have forced significant changes to long standing practices such as landfilling, outdoor incineration, etc. Therefore, due to increasing plastics waste, issues associated with production which have brought concerns about the burden of additional costs in the processing of recycling plastics. When major developments, such as packaging, became widespread and the awareness of the nondegradable nature of plastics waste, it became necessary to engage plastics waste to recycle, reuse and reduce activities [124].

7.23.1 Incineration – Open Air

Incineration of plastics is a threat to the environment. The burning of plastics waste gives off toxic fumes, which could have damaging effects on the population. Toxic fumes stem from the open burning of plastics, a method of incineration that the plastics industry wholly condemns. Plastics waste releases carbon dioxide, which is not only harmful, but also has black smoke which is totally consumed inside a properly operated incinerator.

Traditional incineration of plastics waste is combustion without any control. It releases toxic greenhouse gases. Incineration of these wastes in a traditional way completely destroys all its organic matter. The organic matter present in plastics waste could be valuable for different applications.

In many countries, open air incineration is still carried out due to the sorting problem of continuous dumping at one site. Burning municipal waste including plastics creates atmospheric pollution. While burning, the mixed waste liberates a mixture of gas with high-density smoke and poisonous gases like dioxin. However, incineration has been widely assessed as ecologically unacceptable in the last decade as many environmental regulations have been implemented for a more sustainable recycling-oriented society [125]. Gas liberated from waste burning replaces the oxygen in that area, causing a lack of oxygen. Due to oxygen being unavailable, blood cells struggle due to the lack of oxygen for metabolism, which particularly leads to many diseases, including cancer [126].

7.23.2 Plastics Waste in Concrete

Even though other waste materials, such as fly ash, silica fumes, etc., significantly enhance the properties of concrete due to its hydrophilic nature, with plastics waste powder (PWP), resistivity increases upon degree of saturation. In cement with a lower amount of water there is higher resistivity for the same type of cement due to the hydrophobic nature of plastics waste powder. The degree of contact between the PWP and concrete interfaces can significantly influence the resistivity. Even with pressure, the extent of pressure upon contact and the non-miscibility of plastics waste interface with the concrete to influence resistivity [127]. The density of the concrete with polyethylene terephthalate waste decreases with a decrease in mechanical properties such as compressive strength, as a substitute for sand [128]. Blends of plastics waste and concrete can be used in heavy masses of concreting in pavements. However, the debris with plastics becomes non-separable. The life cycle of waste plastics will not be complete with respect to the environment.

Plastic is one component of municipal solid waste (MSW) which is becoming a major research issue for its possible use in concrete, especially in self-compacting concrete and lightweight concrete. Although some of these materials can be beneficially incorporated into concrete, both as part of the cementitious binder phase or as aggregates, it is important to realize that not all waste materials are suitable for such use and the time or generation burden remains to be answered in the future.

7.23.3 Plastics Waste in Tar for Road Laying

A great number of tar roads are laid and repaired annually. The plastic ground aggregate reused as a partial replacement in road laying could have a deleterious effect, resulting in weakness and cracks affecting the durability. The immiscible nature of tar incorporating plastics waste as a method of destroying plastics is not a permanent solution for future generation from an environmental point of view. The waste remains in tar and becomes microparticles due to degradation. These microparticles damage the nature of the soil, causing problems and becoming a permanent burden on the environment.

However, after waste incineration with energy recovery, the ash that is left over is an excellent valorization route that has the additional environmental advantage of demanding low or no energy for the production of ash materials compared with ordinary Portland cement. There is thus a drive for additional ash materials from waste plastics after incineration that can partially replace cement or tar for use in concrete or tar with ash left over from energy recovery like fly ash from thermal power plants [129].

7.24 Plastics Waste as Blends and Composites

A blend can be technologically compatible if the two plastics are similar in respect to their intermolecular forces or if they interact or react at the phase interface. Compatibilization may be able to reduce the enthalpy of mixing either by making the phase interfaces similar to each other or by providing specific interaction between the phases using additives. This can be achieved by mechanical, chemical, reactive and non-reactive methods [130].

- Mechanical compatibilization involves the adjustment of the viscosity so that the ratios of the viscosity to volume fraction of each component are close to each other and mixing under high shear.
- Chemical compatibilization involves the addition of a small amount of a dispersing agent to help the process by reducing interfacial tension and preventing agglomeration when cooled.

Plastics composite is highly acceptable in aerospace, car racing and the production of sports articles, but have not gained as much ground in mass and serial production areas, such as automobiles, as had been hoped by plastics producers. Composites are attractive materials due to their being manufactured only from plastics waste with sustainable and clean recycling. It is possible to recycle plastics obtained from waste to produce new polymer composites or products, and it can be concluded that plastics waste can be reused to produce useful products or various kinds of composites with the matrix of elastomers or thermoplastics for different applications. Mechanical properties and performance of plastic composite products obtained from plastics waste or recycled plastics depend on the type of matrix, nature of interactions between the blends, and the amount of additives present in it.

7.25 Remarks

Due to its non-biodegradable nature, recycling plastics waste is not a naturally occurring process. It should be viewed as a plastic material just like any other material, which will become waste after usage. However, the cost effectiveness assessed in comparison with other measures should always be considered. Plastics waste presently poses difficulties when addressing final post-use end applications. Using recycling technology does not mean that the technology should be low cost, low maintenance, low level and low effort processing. The success of recycling technology on plastics waste depends on the continued high participation of society and other waste generators. Sociological, institutional and technological advances are needed in order to increase participation and yields in plastics waste recycling [131].

References

1. Hahladakis, J.N., Velis, C.A., Weber, R., Iacovidou, E., Purnell, P., An overview of chemical additives present in plastics: Migration, release, fate and environmental impact during their use, disposal and recycling. *J. Hazard. Mater.*, 344, 179–199, 2018.
2. Marshall, R.E. and Farahbakhsh, K., Systems approaches to integrated solid waste management in developing countries. *Waste Manag.*, 33, 4, 988–1003, 2013.
3. Kyrikou, I. and Briassoulis, D., Biodegradation of Agricultural Plastic Films: A Critical Review, *J. Polym. Environ.*, 15, 125–150, 2007.
4. Verma, R., Vinoda, K., Papireddy, M., Gowda, A., Toxic Pollutants from Plastic Waste- A Review. *Procedia Environ. Sci.*, 35, 701–708, 2016.
5. Grossman D., Hudson J.F., Marks D.H., Waste Generation Models for Solid Waste Collection, in ASCE. *J. Environ. Eng. Div.*, 100, 6 1219–1230, 1974.
6. Wertz, K.L., Economic factors influencing households' production of refuse. *J. Environ. Econ. Technol.*, 2, 263, 1976.
7. Zubero, M.B., Aurrekoetxea, J.J., Ibarluzea, J.M., Arenaza, M.J., Rodriguez, C., Sáenz, J.R., Heavy metal levels (Pb, Cd, Cr and Hg) in

the adult general population near an urban solid waste incinerator. *Science of The Total Environment*, 408, 20, 4468–4474, 2010.
8. Gallardo, A., Carlos, M., Colomer, F.J., Edo-Alcón, N., Analysis of the waste selective collection at drop-off systems: Case study including the income level and the seasonal variation. *Waste Management & Research*, 36, 1 30–38, 2017.
9. Ajayi, S.O. and Oyedele, L.O., Policy imperatives for diverting construction waste from landfill: Experts' recommendations for UK policy expansion. *J. Clean. Prod.*, 147, 57–65, 2017.
10. Ngoc, U.N. and Schnitzer, H., Sustainable solutions for solid waste management in Southeast Asian countries. *Waste Management*, 29, 1982–1995, 2009.
11. Nakamura, S., Kondo, Y., Kagawa, S., Matsubae, K., Nakajima, K., Nagasaka, T., MaTrace: Tracing the Fate of Materials over Time and Across Products in Open-Loop Recycling. *Environmental Science & Technology*, 48, 13, 7207–7214, 2014.
12. Young, C.Y., Ni, S.P., Fan, K-S., Working towards a zero waste environment in Taiwan. *Waste Management & Research*, 28, 3, 236–244, 2009.
13. Stamp, A., Wäger, P.A., Hellweg, S., Linking energy scenarios with metal demand modeling–The case of indium in CIGS solar cells, *Resour. Conserv. Recyc.*, 93, Supp. C, 156–167, 2014.
14. Lennart, Y.L., Materials selection and design for development of sustainable products. *J. Mater. Des.*, 28, 466–79, 2007.
15. Al-Salem, S.M., Lettieri, P., Baeyens, J., Recycling and recovery routes of plastic solid waste (PSW): A review. *Waste Manag.*, 29, 2625–43, 2009.
16. Al-Salem, S.M., Lettieri, P., Baeyens, J., The valorization of plastic solid waste (PSW) by primary to quaternary routes: From re-use energy and chemicals. *Prog. Energy Combust. Sci.*, 36, 103–29, 2010.
17. Ali, M.F. and Siddiqui, M.N., Thermal and catalytic decomposition behavior of PVC mixed plastic waste with petroleum residue. *J. Anal. Appl. Pyrol.*, 74, 282–9, 2005.
18. Moffitt, T.W., *Proceedings of the Industrial Conference on Waste Management in a Green Environment: Developing Technologies and Enhancing Markets for Recyclable Materials*, London, April, 1990.
19. Vanness, K.E. and Nosker, T.J., Commingled Plastics in Plastics Recycling: Products and Processes, R.J. Ehrig (Ed.), Carl Hanser Verlag: Munich, P187–229, 1992.
20. Hamid, S.H. and Muhammad, A., Role of Blending Technology in Polyethylene Recycling. *Journal of Macromolecular Science Part C Polymer Reviews*, 35, 3, 495–515, 1995.

21. Brown, D.T., Chapter1, 1-35, The legacy of the landfill: Perspective on the solid waste crisis, in *Plastics Waste Management-Disposal, Recycling and Reuse*, Mustafa, N. (Ed.), Marcel Dekker, Inc. New York, 1993.
22. Karlsson, S., Recycled Polyolefins. Material Properties and Means for Quality Determination. *Adv. Polym. Sci.*, 169, 201, 2004.
23. Mączka, T., Śliwka, E., Wnukowski, M., Plasma gasification of waste plastics. *J. Ecol. Eng.*, 14, 1, 33–39, 2013.
24. Ha, M.Y., Jeon, C.H., Choi, D.S., Choi, H.J., A Numerical Study on the Triboelectrostatic Separation of PVC Materials from Mixed Plastics for Waste Plastic Recycling. *KSME Int. J.*, 17, 10, 1485–1495, 2003.
25. Steuteville, R., MSW composting at the crossroads. *Biocycle*, 36, 3, 1995.
26. La Mantia, F.P., Perrone, C., Bellio, E., *Recycling of Plastic Materials*, Chem Tech Publishing, Toronto, Canada, pp. 83–98, 1993.
27. Jambeck, J.R., Geyer, R., Wilcox, C., Siegler, T.R., Perryman, M., Andrady, A., Narayan, R., Law, K.L., Marine pollution. Plast. Waste Inputs Land in to Ocean. *Science*, 347, 6223, 768–771, 2015.
28. Leidner, J., *Plastics Waste: Recovery of Economic Value*, Marcel Dekker, Inc. New York, pp. 64–65, 1981.
29. Nhamo, G., Regulating Plastics Waste, Stakeholder Engagement and Sustainability Challenges in South Africa, Urban Forum 19, 83–101, 2008.
30. Farrissey, W.J., Chapter 10, Thermosets in Plastics Processing: Products and Processes, R.J. Ehrig (Ed)., Carl Hanser Verlag: Munich pp. 233–262, 1992.
31. Papaspyrides, C.D. and Poulakis, J.G., In: *Polymeric materials encyclopedia, twelve volume set: Recycling, plastics*. New York: CRC Press Inc.; p. 7403, 1996.
32. Aurrekoetxea, J., Sarrionanlia, M.A., Urrutibeascoa, I., Maspoch, M.L., Effects of recycling on the microstructure and the mechanical properties of isotactic polypropylene. *J. Mater. Sci.*, 36, 2607–13, 2001.
33. Al-Salem, S.M., Lettieri, P., Baeyens, J., Recycling and recovery routes of plastic solid waste (PSW): A review. *Waste Manag.*, 29, 2625–43, 2009.
34. Al-Salem, S.M., Lettieri, P., Baeyens, J., The valorization of plastic solid waste(PSW) by primary to quaternary routes: From re-use energy and chemicals. *Prog. Energy Combust. Sci.*, 36, 103–29, 2010.
35. Bevis, M., Secondary recycling of plastics. *J. Mater. Des.*, 3, 344–9, 1982.

36. Kartalis, C.N., Papaspyrides, C.D., Pfaendner, R., Hoffmann, K., Herbst, H., Mechanical recycling of post-used HDPE crates using the restabilization technique. II: Influence of artificial weathering. *J. Appl. Polym. Sci.*, 77, 1118–27, 2000.
37. Finnveden, G., Johansson, J., Lind, P., Moberg, A., Life Cycle Assessment of Energy from Solid Waste-Part 1: General Methodology and Results, *J. Clean. Prod.*, 13, 213, 2005.
38. Björkdahl, J. and Linder, M., Formulating problems for commercializing new technologies: The case of environmental innovation. *Scandinavian Journal of Management*, 31, 1, 14–24, 2015.
39. Hinsken, H., Moss, S., Pauquet, J.-R., Zweifel, H., Degradation of polyolefins during melt processing. *Polym. Degrad. Stab.*, 34, 279, 1991.
40. Bisio, A.L. and Xanthos, M., "Pyrolytic Processes", Chapter 12, in: "*How to Manage Plastics Waste: Technology and Market Opportunities*", L. Bisio and M. Xanthos (Eds.), pp. 125–143, Carl Hanser Verlag, Munich, New York, 1994.
41. Stangenberg, F., Ågren, S., Karlsson, S., Quality Assessments of Recycled Plastics by Spectroscopy and Chromatography, *Chromatographia*, 59, 101, 2004.
42. Schnabel, W., *Polymer Degradation Principles and practical Applications*, pp. 192–199. MacMillan Publ. Co., New York, 1981.
43. Curlee T. R., and Sujit.D., Viability of recycling plastics by tertiary processes, *Journal of solid waste technology and management*, 25, 3 & 4, 148–160, 1998.
44. Ali, M.F. and Siddiqui, M.N., Thermal and catalytic decomposition behavior of PVC mixed plastic waste with petroleum residue. *J. Anal. Appl. Pyrol.*, 74, 282–9, 2005.
45. Levine, S.E. and Broadbelt, L.J., Detailed mechanistic modeling of high-density polyethylene pyrolysis: Low molecular weight product evolution. *Polym. Degrad. Stab.*, 94, 5, 810–822, 2009.
46. Sinn, H., Kaminsky, W., Janning, J., Processing of Plastic Waste and Scrap Tires into Chemical Raw Materials, Especially by Pyrolysis. *Angew. Chem. Int. Ed. Engl.*, 15, 660, 1976.
47. Coilin, G., Grigoleit, G., Bracker, G.-P., Pyrolytische Rohstoff-Rückgewinnung aus Sonderabfiillen. *Chem. Ing. Tech.*, 50, 836, 1978.
48. Kaminsky, W., Pyrolysis of plastic waste and scrap tyres in a fluid bed reactor. *Resource Recovery and Conservation*, 5, 205, 1980.
49. Kaminsky, W. and Sinn, H., Aromatics from scrap tires. *Hydrocarbon Process.*, 59, 187, 1980.
50. Onwudili, J.A., Insura, N., Williams, P.T., Composition of products from the pyrolysis of polyethylene and polystyrene in a closed batch

reactor: Effects of temperature and residence time. *J. Anal. Appl. Pyrol.*, 86, 293–303, 2009.
51. Bockhorn, H., Hornung, A., Hornung, U., Stepwise pyrolysis for recycling of plastic mixtures. *Macromol. Symp.*, 135, 35–41, 1998.
52. Kaminsky, W., Menzel, J., Sinn, H., Recycling of plastics. *Conservation & Recycling*, 1, 1, 91–110, 1976.
53. Ilgenfritz, E.M., Plastics waste handling practices in solid waste management, *Water, Air, and Soil Pollution*. 4, 191–199, 1975.
54. Li, A.M., Li, X.D., Li, Y.R., Chi, Y., Yan, J.H., Cen, K.F., Pyrolysis of solid waste in a rotary kiln: Influence of final pyrolysis temperature on the pyrolysis products, *J. Anal. Appl. Pyrol.*, 50, 149–162, 1999.
55. Westerhout, R.W.J., Waanders, J., Kuipers, J.A.M., van Swaaij, W.P.M., Development of a Continuous Rotating Cone Reactor Pilot Plant for the Pyrolysis of Polyethene and Polypropene. *Ind. Eng. Chem. Res.*, 37, 2293–2300, 1998.
56. Manos, G., Garforth, A., Dwyer, J., Catalytic Degradation of High-Density Polyethylene on an Ultrastable-Y Zeolite. Nature of Initial Polymer Reactions, Pattern of Formation of Gas and Liquid Products, and Temperature Effects. *Ind. Eng. Chem. Res.*, 39, 5, 1203, 2000.
57. Diaz-Silvarrey, L. and Phan, A., Kinetic study of municipal plastic waste, *Int. J. Hydrogen Energy*, 41, 16352–16364, 2016.
58. Williams, P.T. and Williams, E.A., Interaction of Plastics in Mixed-Plastics Pyrolysis, *Energy & Fuels*, 13, 1, 188–196, 1999.
59. Williams, E. and Williams, P., Analysis of products derived from the fast pyrolysis of plastic waste. *J. Anal. Appl. Pyrol.*, 40, 347–363, 1997.
60. Pinto, F., Costa, P., Gulyurtlu, I., Cabrita, I., Pyrolysis of plastic wastes. 1. Effect of plastic waste composition on product yield. *J. Anal. Appl. Pyrol.*, 51, 39–55, 1999.
61. Sakata, Y., Uddin, M., Koizumi, K., Murata, K., Thermal degradation of polyethylene mixed with poly(vinyl chloride) and poly(ethyleneterephthalate). *Polym. Degrad. Stab.*, 53, 111–117, 1996.
62. Miskolczi, N., Barthaa, L., Deaka, G., Jover, B., Thermal degradation of municipal plastic waste for production of fuel-like hydrocarbons. *Polym. Degrad. Stab.*, 86, 357–366, 2004.
63. Bhaskar, T., Uddin, M., Murai, K., Kaneko, J., Hamano, K., Kusaba, T., Muto, A., Sakata, Y., Comparison of thermal degradation products from real municipal waste plastic and model mixed plastics. *J. Anal. Appl. Pyrol.*, 70, 579–587, 2003.
64. Adrados, A., de Marco, I., Caballero, B.M., López, A., Laresgoiti, M.F., Torres, A., Pyrolysis of plastic packaging waste: A comparison

of plastic residuals from material recovery facilities with simulated plastic waste. *Waste Management*, 32, 5, 826–832, 2012.
65. Lee, K., Material Recycling-Trends and Perspectives. *InTech*, 175–192, 2012.
66. Mastral, F., Esperanza, E., Garcıa, P., Juste, M., Pyrolysis of high-density polyethylene in a fluidised bed reactor. Influence of the temperature and residence time. *J. Anal. Appl. Pyrol.*, 63, 1–15, 2002.
67. García, A.N. and Font, R., Thermogravimetric kinetic model of the pyrolysis and combustion of ethylene-vinyl acetate copolymer refuse. *Fuel*, 83, 1165–1173, 2004.
68. Sorum, L., Gronli, M.G., Hustad, J.E., Pyrolysis characteristics and kinetics of municipal solid wastes. *Fuel*, 80, 1217–1227, 2001.
69. Kruse, T.M., Woo, O.S., Broadbelt, L.J., Detailed mechanistic modelling of polymer degradation: Application to polystyrene. *Chem. Eng. Sci.*, 56, 971–979, 2001.
70. Beyler, C., Thermal decomposition of polymers, Chapter 1-7, in: *The SFPE Handbook of Fire Protection Engineering*, P.J. DiNenno (Ed.), p. 110–130, NFPA, Quincy, MA, (Section 1, Chapter 12), 1988.
71. de Marco, I., Caballero, B.M., López, A., Laresgoiti, M.F., Torres, A., Chomón, M.J., Pyrolysis of the rejects of a waste packaging separation and classification plant. *J. Anal. Appl. Pyrol.*, 85, 384–91, 2009.
72. Garforth, A.A., Lin, Y.H., Sharratt, P.N., Dwyer, J., Lin, Y-H., Sharratt, P.N., J Dwyer Production of hydrocarbons by catalytic degradation of high density polyethylene in a laboratory fluidised-bed reactor. *Applied Catalysis A: General*, 169, 2, 331–342, 1998.
73. Escola, J.M., Aguado, J., Serrano, D.P., García, A., Peral, A., Briones, L., Calvo, R., Fernandez, E., Catalytic hydroreforming of the polyethylene thermal cracking oil over Ni supported hierarchical zeolites and mesostructured aluminosilicates. *Appl. Catal. B Environ.*, 106, 405–15, 2011.
74. Renzini, M.S., Lerici, L.C., Sedran, U., Pierella, L.B., Stability of ZSM-11 and BETA zeolites during the catalytic cracking of low-density polyethylene. *J. Anal. Appl. Pyrol.*, 92, 450–5, 2011.
75. Elordi, G., Olazar, M., Aguado, R., Lopez, G., Arabiourrutia, M., Bilbao, J., Catalytic pyrolysis of high density polyethylene in a conical spouted bed reactor. *J. Anal. Appl. Pyrol.*, 79, 450–5, 2007.
76. Jung, S.-H., Cho, M.-H., Kang, B.-S., Kim, J.-S., Pyrolysis of a fraction of waste polypropylene and polyethylene for the recovery of BTX aromatics using a fluidized bed reactor. *Fuel Process. Technol.*, 91, 277–84, 2010.
77. Iliopoulou, E.F., Stefanidis, S., Kalogiannis, K., Psarras, A.C., Delimitis, A., Triantafyllidis, K.S. *et al.*, Pilot-scale validation of Co-ZSM-5

catalyst performance in the catalytic upgrading of biomass pyrolysis vapours. *Green Chem.*, 16, 662–74, 2014.

78. Yuan, G., Chen, D., Yin, L., Wang, Z., Zhao, L., Wang, J.Y., High efficiency chlorine removal from polyvinyl chloride (PVC) pyrolysis with a gas–liquid fluidized bed reactor. *Waste Manag.*, 34, 1045–50, 2014.
79. Ng, S.H., Conversion of polyethylene blended with vgo to transportation fuels by catalytic cracking. *Energy Fuels*, 9, 216–24, 1995.
80. Elordi, G., Olazar, M., Lopez, G., Amutio, M., Artetxe, M., Aguado, R. *et al.*, Catalytic pyrolysis of HDPE in continuous mode over zeolite catalysts in a conical spouted bed reactor. *J. Anal. Appl. Pyrol.*, 85, 345–51, 2009.
81. Arabiourrutia, M., Elordi, G., Lopez, G., Borsella, E., Bilbao, J., Olazar, M., Characterization of the waxes obtained by the pyrolysis of polyolefin plastics in a conical spouted bed reactor. *J. Anal. Appl. Pyrol.*, 94, 230–7, 2012.
82. Artetxe, M., Lopez, G., Elordi, G., Amutio, M., Bilbao, J., Olazar, M., Production of light olefins from polyethylene in a two-step process: Pyrolysis in a conical spouted bed and downstream high-temperature thermal cracking. *Ind. Eng. Chem. Res.*, 51, 13915–23, 2012.
83. Artetxe, M., Lopez, G., Amutio, M., Elordi, G., Bilbao, J., Olazar, M., Cracking of high density polyethylene pyrolysis waxes on HZSM-5 catalysts of different acidity. *Ind. Eng. Chem. Res.*, 52, 10637–45, 2013.
84. Hussain, Z., Khan, K.M., Perveen, S., Hussain, K., Voelter, W., The conversion of waste polystyrene into useful hydrocarbons by microwave–metal interaction pyrolysis. *Fuel Process. Technol.*, 94, 145–50, 2012.
85. Ludlow-Palafox, C. and Chase, H.A., Microwave-induced pyrolysis of plastic wastes. *Ind. Eng. Chem. Res.*, 40, 4749–56, 2001.
86. Huang, W.-C., Huang, M.-S., Huang, C.-F., Chen, C.-C., Ou, K.-L., Thermochemical conversion of polymer wastes into hydrocarbon fuels over various fluidizing cracking catalysts. *Fuel*, 89, 2305–16, 2010.
87. Yuana, G., Chena, D., Yina, L., Wang, Z., Zhaob, L., Wang, J.Y., *High efficiency of HCl removal from waste plastics pyrolysis with a gas fluidized reactor*. The 1st IWWG-ARB Symposium Hokkaido University Japan, 2013.
88. Undri, A., Rosi, L., Frediani, M., Frediani, P., Efficient disposal of waste polyolefins through microwave assisted pyrolysis. *Fuel*, 116, 662–671, 2014.
89. Hussain, Z., Khan, K.M., Basheer, N., Hussain, K., Co-liquefaction of Makarwal coal and waste polystyrene by microwave–metal interaction pyrolysis in copper coil reactor. *J. Anal. Appl. Pyrol.*, 90, 53–5, 2011.

90. Lam, S.S. and Chase, H.A., A review on waste to energy processes using microwave pyrolysis. *Energies*, 5, 4209–32, 2012.
91. Bai, F., Zhu, C.-C., Liu, Y., Yuan, P.-Q., Cheng, Z.-M., Yuan, W.-K., Co-pyrolysis of residual oil and polyethylene in sub- and supercritical water. *Fuel Process. Technol.*, 106, 267–74, 2013.
92. Goto, M., Chemical recycling of plastics using sub- and supercritical fluids. *J. Supercrit. Fluids*, 47, 500–7, 2009.
93. Tan, X.-C., Zhu, C.-C., Liu, Q.-K., Ma, T.-Y., Yuan, P.-Q., Cheng, Z.-M. *et al.*, Co-pyrolysis of heavy oil and low density polyethylene in the presence of supercritical water: The suppression of coke formation. *Fuel Process. Technol.*, 118, 49–54, 2014.
94. Su, X., Zhao, Y., Zhang, R., Bi, J., Investigation on degradation of polyethylene to oils in supercritical water. *Fuel Process. Technol.*, 85, 1249–58, 2004.
95. Wang, Y. and Zhang, F.-S., Degradation of brominated flame retardant in computer housing plastic by supercritical fluids. *J. Hazard. Mater.*, 205, 156–63, 2012.
96. Onwudili, J.A. and Williams, P.T., Degradation of brominated flame-retarded plastics (Br-ABS and Br-HIPS) in supercritical water. *J. Supercrit. Fluids*, 49, 356–68, 2009.
97. Lilac, W.D. and Lee, S., Kinetics and mechanisms of styrene monomer recovery from waste polystyrene by supercritical water partial oxidation. *Adv. Environ. Res.*, 6, 9–16, 2001.
98. Ng, S.H., Seoud, H., Stanciulescu, M., Sugimoto, Y., Conversion of polyethylene to transportation fuels through pyrolysis and catalytic cracking. *Energy Fuels*, 9, 735–42, 1995.
99. Arandes, J.M., Abajo, I., López-Valerio, D., Fernández, I., Azkoiti, M.J., Olazar, M. *et al.*, Transformation of several plastic wastes into fuels by catalytic cracking. *Ind. Eng. Chem. Res.*, 36, 4523–9, 1997.
100. Papaspyrides, C.D., Poulakis, J.G., Varelides, P.C., A model recycling process for low density polyethylene. *Resour. Conserv. Recycl.*, 12, 177–84, 1994.
101. Vivero, L., Barriocanal, C., Alvarez, R., Diez, M.A., Effects of plastic wastes on coal pyrolysis behaviour and the structure of semicokes. *J. Anal. Appl. Pyrol.*, 74, 327, 2005.
102. Marin, N., Collura, S., Sharypov, V.I., Beregovtsova, N.G., Baryshnikov, S.V., Kutnetzov, B.N., Cebolla, V., Weber, J.V., Copyrolysis of wood biomass and synthetic polymers mixtures. Part II: Characterisation of the liquid phases. *J. Anal. Appl. Pyrol.*, 65, 41, 2002.
103. Sharypov, V.I., Beregovtsova, N.G., Kuznetsov, B.N., Membrado, L., Cebolla, V.L., Marin, N., Weber, J.V., Co-pyrolysis of wood biomass

and synthetic polymers mixtures. Part III: Characterisation of heavy products. *J. Anal. Appl. Pyrol.*, 67, 325, 2003.
104. Ishaq, M., Ahmad, I., Shakirullah, M., Khan, M.A., Rehman, H., Bahader, A., Pyrolysis of some whole plastics and plastics–coal mixtures. *Energy Convers. Manage.*, 47, 3216, 2006.
105. Lavric, E.D., Konnov, A.A., De Ruyck, J., Dioxin levels in wood combustion - A review. *Biomass Bioenergy*, 26, 2, 115–145, 2004.
106. Arena, U., Process and technological aspects of municipal solid waste gasification. A review. *Waste Manag.*, 32, 625–639, 2012.
107. Farraji, H., Zaman, N.Q., Vakili, M.Q., Faraji, H.Q., Overpopulation and Sustainable Waste Management, *International Journal of Sustainable Economies Management*, 5, 3, 13–36, 2016.
108. Ram, A., Narkis, M., Kost, J., Reuse of plastics from solid wastes. *Polym. Eng. Sci.*, 17, 4, 274, 1977.
109. Wenguang, M. and Mantia, F.P.L., Processing and mechanical properties of recycled PVC and of homopolymer blends with virgin PVC. *J. Appl. Polym. Sci.*, 59, 759, 1996.
110. Wong, S.L., Ngadi, N., Abdullah, T.A.T., Inuwa, I.M., Current state and future prospects of plastic waste as source of fuel: A review. *Renewable Sust. Energy Rev.*, 50, 1167–1180, 2015.
111. Achilias, D.S. and Karayannidis, G.P., *Proceedings of the Protection and Restoration of Environment VI*, Skiathos, Greece, 1–5th July, p. 925, 2002.
112. Spychaj, T. and Paszun, D., New trends in chemical recycling of poly(ethylene terephthalate), *Macromolecular Symposia*, 135, 1, 137–145, 1998.
113. Lotz, R., Wick, G., Neuhaus, C., 1967. Inventors; Glantzstoff AG, assignee; US3321510. Process for the recovery of dimethyl terephthalate from polyethylene terephthalate.
114. Marathe, M.N., Dabholkar, D.A., Jain, M.K., 1980. Inventors; Sir Padampat Research Centre, assignee; GB2041916. Process for the recovery of dimethyl terephthalate from polyethylene terephthalate polymer waste, 1979.
115. Michel, R.E., 1992. Inventor; Du Pont, assignee; EP484963. Recovery of methyl esters of aromatic acid and glycols from thermoplastic polyesters scrap using methanol vapor.
116. Socrate, C. and Vosa, R., 1995. Inventors; Montefibre, assignee; Continuous process for the recovery of terephthalic acid from waste or used polyalkylene terephthalate polymers, EP662466.
117. Viksne, A., Kalnins, M., Rence, L., Berzina, R., Unsaturated polyester resin based on PET waste products from glycolysis by ethylene,

propylene, and diethylene glycols and their mixtures. *Arabian J. Sci. Eng.*, 27, 33, 2002.
118. Mansour, S.H., Ikladious, N.E., Depolymerization of poly(ethylene terephthalate) wastes using 1,4-butanediol and triethylene glycol. *Polym. Testing*, 21, 497, 2002.
119. Paszun, D. and Spychaj, T., Chemical Recycling of Poly(ethylene terephthalate). *Ind. Eng. Chem. Res.*, 36, 1373, 1997.
120. Karayannidis, G.P., Achilias, D.S., Sideridou, I.D., Bikiaris, D.N., Alkyd resins derived from glycolized waste poly(ethylene terephthalate). *Eur. Polym. J.*, 41, 201, 2005.
121. Karayannidis, G.P., Nikolaidis, A.K., Sideridou, I.D., Bikiaris, D.N., Achilias, D.S., Chemical Recycling of PET by Glycolysis: Polymerization and Characterization of the Dimethacrylated Glycolysate. *Macromolecular Materials and Engineering*, 291, 11, 1338–1347, 2006.
122. K.P. Blackmon, D.W. Fox, S.J. Shafer, Process for Converting PET Scrap to Diamide Monomers. Eur. Patent 365 842, 1988.
123. Cai, J-j., Yu, G-w., Liao, H-q., Qian, K., Zhao, P., He, Y-b., Disposal of Waste Plastics With Traditional Coking Process, *Journal of Iron and Steel Research, International*, 13, 1, 05–09, 2006.
124. Medina, M., The effect of income on municipal solid waste generation rates for countries of varying levels of economic development: A model. *Journal of Resource Management and Technology*, 24, 149–155, 1997.
125. Achilias, D.S., Chemical Recycling of Polymers. The Case of Poly(methyl methacrylate). Proceedings of the 2006 IASME/WSEAS International Conference on Energy & Environmental Systems, Chalkida, Greece, May 8–10, 271–276, 2006
126. Studies on air pollution with plastics –Dr. Muralisrinivasan Natamai Subramanian – unpublished.
127. Polder, R.B., Test methods for on site measurement of resistivity of concrete—A RILEM TC-154 Technical Recommendation. *Constr. Build. Mater.*, 15, 125–131, 2001.
128. Marzouk, Y., Dheilly, R.M., Queneudec, M., Valorization of post-consumer waste plastic in cementitious concrete composites. *Waste Technol.*, 27, 2, 310–318, 2007.
129. Studies on micro-particles in road debris – unpublished report – Muralisrinivasan Natamai Subramanian.
130. Markham, R.L., Ch. 2, in: *Elastomer Technology Handbook*, P. Cheremisinoff (Ed.), CRC Press, Boca Raton, FL, 2005.
131. Kovacs, W.L., The Coming Era of Conservation and Industrial Utilization of Recyclable Materials. *Ecol. Law Q.*, 15, 537, 1988.

8
Economy and Recycle Market

Plastics have a number of characteristics which are superior to other materials in specific areas. The properties include resilience to environmental influence, stiffness, light weight, and the ability to produce complex geometry forms with relative ease. By virtue of these properties, products made of plastic have driven other products out of the market.

The plastics processing industry is one of the few branches in which output has strongly expanded. The propensity to invest is also higher than in the manufacturing industry as a whole, albeit against the background of plastics recycling. The production of plastic goods is labor intensive. Consumption focuses largely upon the mass production of polyvinylchloride (PVC), polyethylene (PE), polystyrene (PS) and polypropylene (PP) products. Plastics are deployed in a broad spectrum of fields of application. The most important area that uses plastics is the construction industry, followed by the packaging sector and electrical and electronics industry.

8.1 Economical Background

The plastics processing industry is a relatively young industry which achieved its breakthrough in the 1950s and 1960s with the rapid development of petrochemicals. Since then plastics waste has seen a continuous rise in use, with an increase in the number of different plastics which have displaced more traditional materials in a large number of fields and applications. New products and markets have been created on the basis of plastics. From the beginning of the 1990s till now, the plastics industry continues to be among the fastest growing branches of the world economy.

Muralisrinivasan Natamai Subramanian (ed.) Plastics Waste Management, 2nd Edition (163–182)

8.2 Growth Trajectory

Over the last twenty years, ever since industries embarked on a growth trajectory, plastic materials have been an economic miracle. Macroeconomic statistics, such as measures of economic growth, price change, and incomes, have drawn increasing scrutiny from economists and sociologists seeking to understand the origins of plastics waste and their rise to prominence in existing raw material and their manufacturing units. Now and in the future, environmental protection is an important area at the global level. Although environmental protection legislation is initially perceived by companies as imposing additional costs, it can nevertheless prove advantageous if it induces solutions that strengthen the firms' competitive position in the long run. Advanced economies must set standards that ensure that economic growth is sustainable in social and environmental terms. Europe, as a highly concentrated economic area, can play a leading role for similar markets. This is especially true for the Asian markets, which, once economic growth picks up again, will quickly encounter constraints if they do not come to grips with the problem of waste. For this reason, industry in general, and plastics processing in particular, is well advised to prepare themselves to make full use of the systemic advantages offered by integrated environmental protection in Europe and particularly in the USA.

8.3 Value of Plastics Waste

Various factors drive the demand for the recovery of material from industrial waste. The economic value of the plastics industry relies on virgin and recycled material, the technology of using virgin material, the substitution of recycled material for virgin material and the market structure. The lack of record-keeping within industrial sectors presents a major challenge for the verification of the actual amounts of materials that are eventually recycled; knowledge of the energy saved as a result of the reuse of recycled materials is a key factor in the utilization of industrial waste.

Due to the realization valuable plastics waste, there has been increased focus on the collection or separation of plastic waste from other mixed wastes; hence, the recycling of plastics waste is part of a huge, billion dollar market. The pattern of buying dictates the use of resource required and hence prioritizes the type and amount of recycled plastic that is targeted [1]. Recycled plastics are used in small amounts by a multitude of industrial firms all over the world and the waste plastics market is used by many small-scale industries.

Plastics waste adds value when recycled alone or with a certain percentage of virgin material as long as the essential properties of the end product are not negatively impacted. The recycling of waste plastics may have to look beyond its initial application in order to alleviate any detrimental environmental impact. A crucial strategy in the management of plastics waste will be the collection, segregation and effective handling of the whole process from the solid waste stage.

8.4 Economic Issues

Economic and environmentally friendly solutions are the goal in order to reduce dependence on fossil fuels through more efficient energy technologies. All actions have to ensure alternative routes for fuel production and implementation of waste management. An overview of economic issues involved in time limitations on waste plastics recycling to the greatest possible extent is presented. The process includes cleaning, separation, granulation, and processing.

Environmental protection plays an increasing important role. In the area of mass production, materials are recycled underway on concepts for a circular economy. Materials and products must be conceived in possible recycling without loss of quality. Products require desired optimization include entire product life cycle.

Circular economy and zero waste movements have brought more attention to waste prevention initiatives in the management of solid wastes in recent years, increasing awareness of reuse as a means for conservation [Figure 8.1].

The economic impact of plastics has been focused on sustainable production and the decoupling of adverse environmental effects from waste generation, since plastics waste disposal has become a major environmental problem worldwide.

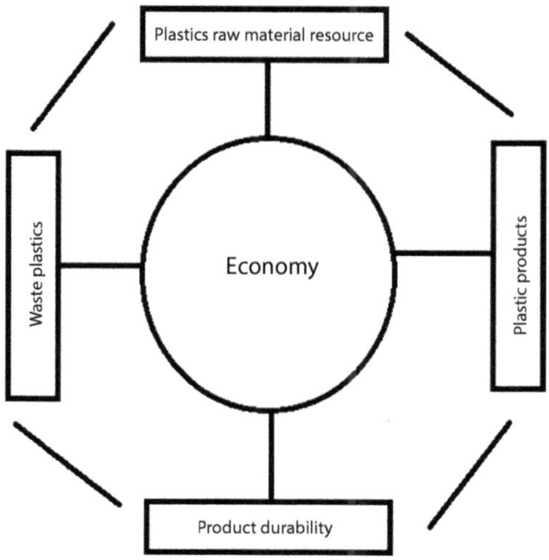

Figure 8.1 Circular economy of waste plastics.

8.5 Market Dynamics and Uncertainty

The rapid development of the market economy of plastics production is accompanied by an increase in production capacity. This leads to an increase in plastics waste, which causes some direct and indirect risks related to environmental pollution. Plastic products made from virgin material are facing increased competition and a depletion of natural sources of raw material. Economists who are critical of classical and neo-classical economics state that an important element in any theory of economics is uncertainty [2]. Plastics waste has become a recognized source of recycled material sales based on market uncertainty.

The recycled plastics market would increase in the future if industries changed their state of ignorance regarding their collective concepts on the use of waste plastics material. Risk taking and innovation through the separation of plastics waste from other solid waste material will characterize the market trend.

The separation of plastics changes the recycling utilization structure and strategies involved, which impacts the development of the recycled plastics market by allowing new opportunities to be sought, errors and constraints to be overcome, and planning to be implemented. This would result in changes within the market place [3] brought about by industries seeking to increase the use of plastics waste in order to reduce cost and increase knowledge, which would result in reducing the market uncertainty with regards to non-degradable plastics waste. In order to survive, industries must expand their knowledge, which requires them to enter into networks of competition with key players in their sales and marketing chain. The future of the plastics waste market is unknown as it is characterized by high levels of competition between industries.

Market share of technological application or market penetration is related to the recycled plastics market share of technologies that use the specific resource.

- Concentration of recycled plastics supply is related to the change in the waste plastics supply dynamics from different regions.
- Price of a commodity can influence and be influenced by the waste plastics supply and demand of a resource as well as by speculative anticipation.

Due to increased competition, cost reduction and economic uncertainty, industry is seeking to manufacture products from a combination of recycled and virgin material. The competitive situation in world markets has led to a move away from the use of pure virgin material; hence, the time has come to use plastics waste in large quantities. The role of marketing plastics waste involves finding new strategies and infrastructures for novel market opportunities.

8.6 Fiscal Waste

The problem of wastes accumulation in lands contribute environmental problems. Many of these wastes are left as stockpiles or dumped in rivers, oceans and inlands. Pollution is more insidious in developing and developed countries, The waste materials are either buried or incinerated in many countries. Therefore, the protection of the environment has to be taken into account in the process of economic and social development.

Due to the rapidly increasing growth of the plastics industry, a large quantity of plastic products become obsolete every year; therefore, outdated plastic products, including end-of-life plastic products, are the most rapidly growing plastics waste problem worldwide [4]. As a result, many plastic materials, such as polyvinyl chloride, acrylonitrile-butadiene-styrene and so on, along with heavy metals and flame retardants, create serious pollution problems during the recycling and disposal process.

The open burning of waste plastic causes the emission of gases and chemicals which are toxic [5], and the recycling activities of plastics in electronic waste (E-waste) have resulted in high levels of pollution in the ambient environment. These activities threaten the ecosystem [6]; therefore, driven by profit from the increased cost of plastics, primitive waste recycling processes are now extremely active.

8.7 Waste to Value

The use of polyolefin, polystyrene, and vinyl has been increasing much faster than the economy. The concept of energy from plastics waste is being capable of producing fuels and value-added products. The energy recovery technologies under this concept are anaerobic digestion, pyrolysis, transesterification, refuse-derived fuels and incineration.

Industrial Strategy

The industrial strategy is to highlight the importance of competition; intense competition is a result of numerous competitors, slow growth, high-level fixed costs, lack of differentiation amongst products and quality, ease of product substitution and the existence of industries with a high stake in achieving success [7].

Marketing, when properly understood and practiced, is the very soul of enterprise; it gives meaning to and permeates all aspects of it. A business enterprise has two functions: marketing and innovation, the remainder involves cost [8]. If a clear understanding of waste management is critical to success, then a clear understanding of the marketing process is fundamental. Marketing and sales have supplanted manufacturing as the primary aspect of industrial competition. The development of products containing waste plastics has become the battleground as the postindustrial stage begins.

8.8 Industrial Ecology

Rapid economic development has resulted in greater complexity in the management of industrial waste which, in the plastics industry, includes the recycling, reuse and final disposal of unusable material. Industrial waste from the plastics industry contains additives which may be in a concentration high enough to endanger human beings or pollute the environment due to the volume generated. As a result of the additive content of industrial waste, it can be considered hazardous if the extract concentration exceeds the maximum allowed concentration of contaminants. Industries also generate waste other than hazardous industrial waste.

Industrial ecology is a recent concept in engineering and management, and is an attempt to manage an industrial unit as an ecosystem, with feedback loops and the minimal use of resources and production of waste [9]. Life cycle assessment, which was designed to take into account the environmental impact of waste, can help achieve cleaner production technologies and sustainable waste management practices which are all important aspects of industrial ecology. The concept of industrial ecology will help industries to minimize the use of resources and production of waste, from resource extraction to manufacturing, and final consumer use to recycling of the product.

Industrial ecology (IE) aims to study, develop, and implement strategies that include the decrease of environmental impacts of by-products and waste usually associated with industrial systems. It explores various interactions that can occur between industrial systems and the environment. IE adopts a natural paradigm and proposes that industrial ecosystems can behave like a natural ecosystem wherein many resources are used and recycled [10]. To do so, industrial ecosystems need to maximize resource efficiency by sharing their resources and creating opportunities to use or recycle industrial waste and by-products [11–12].

Waste plastics inventory and analysis, there is a continuing controversy about the difference between the production and waste production. Principles of physical organization starts with future waste management system which leads waste collection stations. Furthermore, the degree of collection involves waste to convert energy recovery or material recovery which are both extreme opposites.

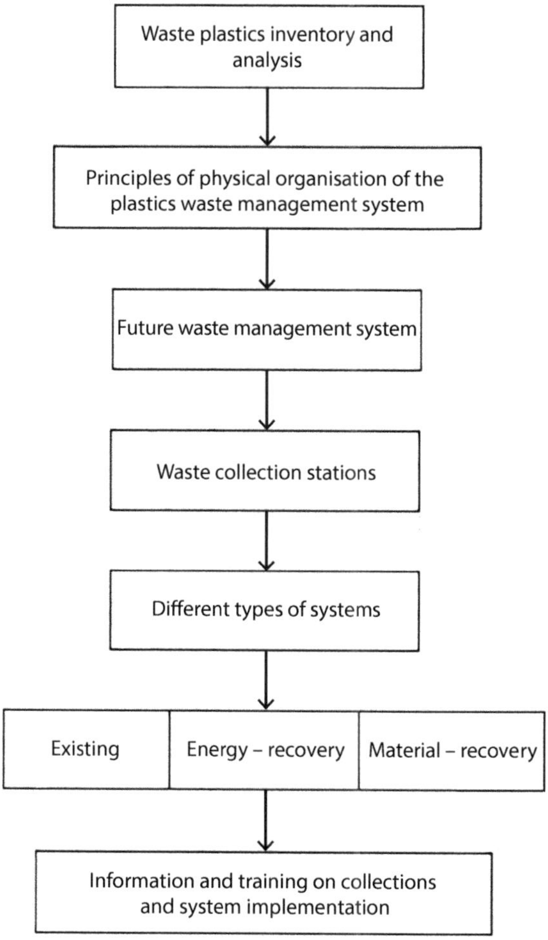

Figure 8.2 Schematic illustration of the waste assessment method.

Therefore, information and training on collections and system implementation [Figure 8.2] is an important to believe in order to have a plastics waste management. Schematic illustration of waste assessment method to entail a unique set of activities or functions.

8.9 Industrial Symbioses (ISs)

Industrial symbioses (ISs) join organizations from different geographic areas and different industrial sectors together in order to share and exchange resources such as energy, materials, and water. Plastics waste can also include information and ideas about business practices that help in sustainable innovation and thus are not limited to material and energy flows [13–15].

Q1

8.10 Economic Advantages

Increasing costs of petroleum products and plastic raw materials underlie the stiking phenomenon of the widening use of recycled plastics. Plastics waste are valuable wastes which are either treated at considerable cost using high-end processes or lose their marketability; they are treated elsewhere by means of scarce resources that do not comply with technical and social standards.

There has been intensive economic analysis of a compatible approach to environmental solutions to improve issues of waste prevention and recycling of waste plastics. The economic success of plastics recycling largely depends on costing and market prices for recycled material [16–26].

8.11 Economic Implications [27, 28]

Incorporating economy into business activities through innovative recycling practices, recovery methods, reuse efficiency and diversification into appropriate reuse methods in plastics recycling, there is sustainable production, processing restraint and reduction in waste generation, and collective imperatives that can lead to significant operational and competitive advantages. However, plastics waste may not have integrated concerns for processing in respective industrial planning processes. With the globalization of world markets, it is important that the processing used for plastics waste is sufficiently flexible to adapt to a variety of business environments. The economics of waste plastics are influenced by

- Problems of poverty and low income
- Existing financial resources are not sufficient for the collection, transportation, appropriate separation and disposal of plastics waste.

- Lack of sufficient sources of investments and absence of state financial resources for investments.
- Political indicators implications
- State, regional and local institutions are under management and jurisdictions of political parties.
- Problems of ongoing problems present in all transition economies.
- Many problems are artificially generated or exaggerated.
- Focus on real problems in the society

8.12 Marketing Strategy

In the marketing of plastics waste, markets grow explosively requiring multinational organizations to develop new strategies to manage the change. However, the changes required in the organizational structure are slow to follow, resulting in people and processes struggling to keep up with manufacturing demand, which has a detrimental impact on the brand image. Today, the acceptance of plastics waste as a useful resource contributes to increasing knowledge of its potential. The analytical research paradigm is not sufficient for investigating complex real-life issues involved in managing the recycling of plastics waste.

Therefore, there is a palpable sense of transformation in both the virgin and recycled product marketing sectors, and rapid changes have been observed in the way recycled product marketing are achieved. An increasing global environmental awareness, has been catalyzed by the sudden end to years of economic growth and some changes outside the realms of the market of recycled products. Rapidly changing and highly complex plastics business environments are affecting the trading process. Marketing is a strategic value-creating process and often attracts high-level management involvement particularly in the field of recycled products. Profit is an important factor for marketing, yet when considering recycling, attention has to be devoted to the budget in terms of cost reduction, processing and so on [29].

8.13 Modern Marketing Philosophy

Recycled plastics marketing, has increasingly been shown to encounter a series of organizational problems to overcome. Therefore, exploration of the relevant concepts in the marketing sector must be embraced if recycled plastics problems are to be faced and overcome [30].

With regards to the market of recycled products, planning and monitoring is linked to expenditure on market activities, emotions, customer demands and future prospects. Remuneration schemes and plans for rewarding increases in the sale of recycled products are common, however, processes and techniques for predicting volume of sale and trends are essential. The marketing of recycled plastics requires familiarity with the concept of profit gap analysis, profit planning and new product development as three interlinked stages which ensure that requisite corporate profitability is attained. Modern marketing philosophy cites profit-seeking activity as opposed to sales volume activity as the distinguishing feature.

Today's competitive landscape has not changed significantly over recent years [31]. Many of the pressures on recycled products seem to be related to changes in customer expectations, which require rapid changes and persistent pressure to achieve the necessary sales quotas [32]. Customer expectations, rising buyer power and fierce competition are all factors in the trading of recycled products.

8.14 Recycled Plastics Market

The plastics waste market will remain in an aimless and confused state, without any clear idea of how to determine technology, and will win or lose based on issues beyond market uncertainty.

Issues associated with increased plastics waste are:

- Liability
- Production loss
- Must be recycled
- Risk element

Plastics waste is likely to become a primary concern of human organizations based on our past and current relationships with plastic products. Although not entirely caused by plastics, numerous large-scale problems that appear to threaten the good health of humans have come to be taken for granted.

If issues related to plastics and increasing disease are not addressed, health care costs can be expected to rise dramatically.

There is a growing interest in increasing the recovery and recycling of waste plastics. The basic arguments regarding productivity activity are [33]:

- An increased willingness to recognize that the raw materials used for the manufacturing of plastics are limited.
- Many major industrial nations are extremely dependent on overseas sources for primary raw materials, which causes potential economic and political disadvantages.
- The social and environmental costs brought about by many cost cutting industrial practices are becoming politically unacceptable.

Plastics waste can vary considerably from industry to industry as does the degree to which recycling can benefit individual industries.

In the field of recycling plastics, the industry faces two major hazards:

1. Failing to take economic advantage offered by new market possibilities.
2. Assuming the recycling market automatically has a profitable market outlet.

Promising trends have been noted relating to marketing strategy being a primary tool in the recycled market, which has highlighted the diminishing levels of natural resources and unavailability of virgin plastic materials. Some promising introductory work has taken place with industries mixing waste plastics in varying proportions with virgin material without affecting product quality; however, the recycling of plastics has received little attention and the sale of recycled products is still at an early stage of development. Customer satisfaction is the key to boost repeat business, customer loyalty, referrals and, as a consequence, increasing long-term profitability.

For marketing and sales of plastic recycled materials research is paramount. It is also necessary for regulatory trends. The marketing and sales of recycled plastics impact the plastics industry. A global market overview tracks developments in waste recycling. Although recycled plastics will not be of the finest quality, the processing cost and investment will be recovered in industry by reusing plastic materials. These types of conversions are game changing developments that redefine the reprocessing industry and its products. It is a pain to achieve a level playing field in the global waste plastics market.

Recycled plastics provide the primary point of differentiation between competitors and as such, they are increasingly critical to the success of many industries and rely on:

- The economic view which suggests that products made using recycled plastics are of acceptable quality and reliability.
- The strategic view considers that the recycling of plastics acts as a mobility barrier to competition.
- The marketing view links recycled plastics with favorable views of product quality.
- Industrial considerations, in terms of mixing recycled plastics with virgin materials.
- The sociological view focuses on social networks and connections in terms of product quality.
- The accounting view considers waste plastics in terms of recycling as a capital asset.

8.15 Industrial Marketing

Industrial marketing of recycled plastics is highly likely to be affected, in business and domestic activities, by the antipollution lobby. Some industries could be adversely affected by the environmental crisis, whilst others could grasp new marketing opportunities resulting from financial support and new legislation.

The recycling market is driven almost exclusively by regulation, and the conditions necessary for stimulating innovation are not unlike those necessary for the virgin plastic markets.

- Majority of the industries will not innovative unless they can appropriate the profits of the innovation for themselves in recycling.
- Industries will have difficulty in assessing the potential profits if recycling market uncertainty clouds their view of the opportunities.

8.16 Product Development and Marketing

As markets move closer together in the global economy and worldwide logistics becomes more accessible, the geographical reach of technologies has increased. If a technology for manufacturing products or services can be offered in various parts of the world this leads to financial benefits for both the producer and consumer.

Product development using plastics waste is a cross-functional process that is largely composed of functional engineering skills and the developing of manufacturing processes. Product development includes many important marketing activities, and to be successful must be closely coupled with integration process. There are many products which are currently under development and are examples of excellent product development processes – which is a considerable accomplishment, but not enough. Additional development of products using waste plastics is urgently needed.

This line of reasoning is limited, however, to cases of products and services where consumer demands are independent of geographical location and cultural context. The field of plastics processing and waste management addressed are very sensitive in principle, this transfer of knowledge appears pertinent, and the success of such an enterprise depends heavily on correct management procedures tailored to the specific location. A growing trend in today's business environment is understanding the value of recycled plastics. Plastics processing industries include an increasing number of firms which offer marginal customization of products and services, which highlights the fact that industries emphasize the importance of production and marketing interactions.

Efforts to translate technical success into commercial success are full of pitfalls. Waste plastics recycling is a billion dollar market in which a few companies sell recycled products to a multitude of industrial firms worldwide. The waste plastics market is voluminous and small firms order this resource in large quantities only when it is needed. Two functional processes that are fundamental and critical for any business are product development and marketing.

Industrial marketing of recycled plastics is highly likely to be affected, in business and domestic activities, by the antipollution lobby. Some industries could be adversely affected by the environmental crisis, whilst others could grasp new marketing opportunities resulting from financial support and new legislation.

The balance between population growth and natural resources is, in a sense, a problem that has always been with us, and also affects all the other species on the planet. Comparatively recent trends of the rapid growth in population, growth in exploitation of the earth's resources and growth in every parameter which characterizes the status of mankind, have led to comprehension of the fact that we live on a finite globe of limited resources. The finiteness of the earth is within sight of imposing the ultimate confrontation between man and nature: between human systems whose influence has now grown worldwide, and the natural ecosystems that have arisen and maintain the biosphere as a place suitable for life [34].

8.17 Recycled Plastic Products and Consumer Market

Consumers are considered rational actors as far as plastic products and their usage are concerned. Industries are assumed to minimize their costs by choosing an optimal combination of reprocessing inputs, which presumes industries always make fully rational choices holding all the relevant information to exhaust all options to maximize the efficiency/reduce the impacts of their supply chains. Households are assumed to behave as a single homogeneous user, acting to maximize their utility by buying the lowest cost plastic products offered on the market to make rational choices. Yet, real-world decision making deviates from the principles of rational choice. Such simplified assumptions by industries have resulted in policies falling short of achieving their technical potential because of unrealistic assumptions about human behavior, leading to waste plastics being an environmental burden [35, 36].

8.18 Remarks

It would have been almost impossible to predict that plastics waste recycling would markedly reduce new savings in response to the general economic picture, consuming a larger part of waste. With a slow economy, prices are increased at the consumer level. With the renewed rise under economic control, the rise in value-added tax rates and charges in the public sector affect the market growth. In addition, weak demand for many products coupled with intense competition among suppliers help to make for relatively stable prices at the cost of quality.

Even though recycled plastics have changed competitive relations with virgin materials, industries processing plastics will be able to use virgin material more easily due to their contaminant-free purity. The recycling market can expand both sales and production only with competitive advantages through innovation.

The rapid development of the market economy is accompanied by the growth of plastic raw materials volumes of provided manufacturers. This leads to the increasing of plastics waste which causes some direct and indirect risks of the environmental pollution. Consumers with distinct levels of economic development will dispose of plastics waste in different ways. It is necessary to break the link between plastics waste and economic growth.

The global market provides an overview of the developments in plastics waste recycling. It is paramount that marketing and sales of recycled plastics is researched. Different disciplines, designing studies by conducting the right experiments, and capturing and reporting the key data are critical aspects of any research program. Rapid, significant, and measurable impact through research translates into a commercial process. It is difficult to achieve a level playing field in the global waste plastics market. Although recycled plastics will not be up to the standards of virgin material, the processing cost and investment will be recovered. By reusing the plastic materials, the processing industry and its products will be redefined.

Compensation may serve to elide the intrinsic links between economic development, affluence and waste from plastics and waste disposal. For every ton of waste, multiple tons of plastics product and packaging are generated by commercial business. Assigning consumers responsibility for plastics waste would be narrow-minded. Therefore, waste disposal practices may be achieved, rather than being sources of environmental issues [37].

Technical expertise, operating efficiency, strong strategic alliances and regulatory technology preferences will be the primary ways in which to capture market share.

References

1. Fox, H.W., The case of the pulled plasterboard. *Industrial Marketing Management*, 3, 315, 1972.
2. Machovec, F.M., Chapter 11, Stylized Assessment of Gain Versus Pain, in: *Perfect Competition and the Transformation of Economics*, Routledge, London, UK, p. 265, 1995.
3. Kirzner, I.M., Entrepreneurial Discovery and the Competitive Market Process: An Austrian Approach. *J. Econ. Lit.*, 35, 60, 1997.
4. Halluite, J., Linton, J.D., Yeomans, J.S., Yoogalingam, R., The challenge of hazardous waste management in a sustainable environment: Insights from electronic recovery laws. *Environ. Manage.*, 12, 31, 2005.
5. Wang, D.L., Cai, Z.W., Jiang, G.B., Leung, A., Wong, M.H., Wong, M.K., Determination of polybrominated diphenyl ethers in soil and sediment from an electronic waste recycling facility. *Chemosphere*, 60, 810, 2005.
6. Zheng, L.K., Wu, K.S., Li, Y., Qi, Z.L., Han, D., Zhang, B., Gu, C.W., Chen, G.J., Liu, J.X., Chen, S.J., Xu, X.J., Huo, X., Blood lead and cadmium levels and relevant factors among children from an e-waste recycling town in China. *Environ. Res.*, 108, 15, 2008.
7. Porter, M.E., Chapter 1, Competitive Strategy: The Core Concepts, in: *Competitive Strategy: Techniques for Analyzing Industries and Competitors*, Free Press, New York, NY, USA, 1980.
8. Drucker, P.F., *People and Performance: The Best of Peter Drucker on Management*, Harper & Row, New York, NY, 1977.

9. Frosch, R.A. and Gallopoulos, N.E., Strategies for manufacturing. *Sci. Am.*, 216, 3, 144–152, 1989.
10. Davis, C., Nikolic, I., Dijkema, G.P., Industrial Ecology 2.0. *Journal of Industrial Ecology*, 14, 5, 707–726, 2010.
11. Davis, C.B., Rivela, B., Janowicz, K., Semantic catalogs for life cycle assessment data. *Journal of Cleaner Production*, 137, 1109–1117, 2016
12. Zhang, Y., Luo, X., Buis, J.J., Sutherland, J.W., LCAoriented semantic representation for the product life cycle. *Journal of Cleaner Production*, 86, 146–162, 2015.
13. Cecelja, F., Raafat, T., Trokanas, N., Innes, S., Smith, M., Yang, A., Kokossis, A. et al. e-Symbiosis: Technology enabled support for industrial symbiosis targeting small and medium enterprises and innovation. *Journal of Cleaner Production*, 98, 336–352, 2015.
14. Grant, G.B., Seager, T.P., Massard, G., Nies, L., Information and communication technology for industrial symbiosis. *Journal of Industrial Ecology*, 14, 5, 740–753, 2010.
15. Walls, J.L. and Paquin, R.L., Organizational perspectives of industrial symbiosis: A review and synthesis. *Organization & Environment* 28, 1, 32–53, 2015.
16. Rosa, P. and Terzi, S., Improving end of life vehicle's management practices: An economic assessment through system dynamics. *J. Clean. Prod.*, 184, 520–536, 2018.
17. Van Eygen, E., Laner, D., Fellner, J., Circular economy of plastic packaging: Current practice and perspectives in Austria. *Waste Manag.*, 72, 55–64, 2018.
18. Rujnic-Sokele, M. and Pilipovic, A., Challenges and opportunities of biodegradable plastics: A mini review. *Waste Manag. Res.*, 35, 132–140, 2017.
19. Wang, H., Gu, Y., Li, L., Liu, T., Wu, Y., Zuo, T., Operating models and development trends in the extended producer responsibility system for waste electrical and electronic equipment. *Resour. Conserv. Recycl.*, 127, 159–167, 2017.
20. Eren Yıldız-Geyhan, Gökçen Alev Altun-Çiftçioğlu, Mehmet Arif Neşet Kadırgan, Social life cycle assessment of different packaging waste collection system. *Resour. Conserv. Recycl.*, 124, 1–12, 2017.
21. Wiesmeth, H. and Häckl, D., Integrated environmental policy: A review of economic analysis. *Waste Manag. Res.*, 35, 4, 332–345, 2017.
22. Markovic, S. and Bagherzadeh, M., How does breadth of external stakeholder co-creation influence innovation performance? Analyzing the

mediating roles of knowledge sharing and product innovation. *J. Bus. Res.*, 88, 173–186, 2018.
23. Ragaert, K., Delva, L., Van Geem, K., Mechanical and chemical recycling of solid plastic waste. *Waste Manag.*, 69, 24–58, 2017.
24. Munari, C., Scoponi, M., Mistri, M., Plastic debris in the Mediterranean Sea: Types, occurrence and distribution along Adriatic shorelines. *Waste Manag.*, 67, 385–391, 2017.
25. Orset, C., Barret, N., Lemaire, A., How consumers of plastic water bottles are responding to environmental policies? *Waste Manag.*, 61, 13–27, 2017.
26. Ragossnig, A.M. and Schneider, D.R., What is the right level of recycling of plastic waste? *Waste Manag. Res.*, 35, 129–131, 2017.
27. Troschinetz, A.M. and Mihelcic, J.R., Sustainable recycling of municipal solid waste in developing countries. *Waste Manag.*, 29, 915–923, 2009.
28. Tsilemou, K. and Panagiotakopoulos, D., Approximate cost functions for solid waste treatment facilities. *Waste Manag. Res.*, 24, 310–322, 2006.
29. Geiger, S. and Guenzi, P., The sales function in the twenty-first century: Where are we and where do we go from here? *Eur. J. Market.*, 43, 7–8, 873, 2009.
30. Wills, G. and Hayhurst, R., Organisational transfer to marketing futures: An analysis of new patterns of marketing activity for successful development in the seventies. *Ind. Market. Manage.*, 1, 47, 1971.
31. McNamara, G., Vaaler, P.M., Devers, C., Same as it ever was: The search for evidence of increasing hypercompetition. *Strat. Manage. J.*, 24, 3, 261, 2003.
32. Ingram, T.N., Future themes in sales and sales management: Complexity, collaboration, and accountability. *J. Market. Theory Pract.*, 12, 4, 18, 2004.
33. Oliver, A.N. and Rothman, H., The value of recycling: A framework for analysis. *Ind. Market. Manage.*, 4, 133, 1975.
34. May, R.M., The environmental crisis: A survey. *Ind. Market. Manage.*, 2, 211, 1972.
35. Scott, K., Giesekam, J., Barrett, J., Owen, A., Bridging the climate mitigation gap with economy-wide material productivity. *J. Ind. Ecol.*, 1–14, 2018.
36. Scott, K., Roelich, K., Owen, A., Barrett, J. *et al.*, Extending European energy efficiency standards to include material use: An analysis. *Clim. Policy*, 18, 5, 627–641, 2018.
37. Cooper, T., Slower Consumption: Reflections on Product Life Spans and the 'Throwaway Society'. *J. Ind. Ecol.*, 9, 1–2, 51–67, 2005.

9
Life Cycle Assessment

Earth is a closed thermodynamic system with respect to matter and energy with clear distinction. Substances such as plastics that are thrown away do not disappear. At best they end up where nothing more can be done with them. The implication is that plastics should be used for as long as possible as recycled material. In this way, plastics waste can combat both the diffuse spread of waste and their accumulation as reusable materials.

Because of the light weight of plastics, human are subject to significant exposure to them as a direct result of the widespread use of various types of many products such as construction materials, automotive parts, medical supplies, household products, electricity and electronic industries, agriculture, toys and packaging. Plastics have become a basic material in almost all fields. An exhaustive list of solid waste includes plastic waste available today from households, industry, and, finally, consumer packaging markets.

9.1 LCA and Plastics Waste

One of the significant problems of state recycling management is to collect recyclable materials and to support sorting activities, it is an excellent way to use the product lifecycle and sustainable reuse to achieve urban waste management [1]. Transportation in economic systems, including services, production, and distribution has a special importance and it accounts for the significant part of gross national product of countries.

Life cycle assessment (LCA) of waste management solutions has come a very long way with the establishment and evaluation of individual technology [2–4] to the current, much more mature level in

Muralisrinivasan Natamai Subramanian (ed.) Plastics Waste Management, 2nd Edition (183–198) © 2019 Scrivener Publishing LLC

which extensive and integrated view of uncertainties, methodological implications, and detailed framework conditions [5, 6].

Plastics waste after separation and purification can be reused in production. Although many methods, such as chemical, thermal, and thermomechanical, have been used for processing plastics waste, these have often been costly and not suitable for mixed plastics waste. A production in which waste plastics are able to flow under pressure and heat is an alternative to mixed plastics waste to produce products at reasonable cost [7–9]. Converting waste plastics to raw material provides a technologically viable alternative that gives them new life instead of landfilling them. Raw material utilization could offer economic advantages and reflect recycling philosophy. Recycling can become part of the life cycle of a product. The numerous elements constituting the realm of plastics waste are defined and presented along with life cycle assessment. These includes domestic waste, E-waste, etc. The life cycle assessment tools used today in research and industry is to be analyzed.

Background

The first life cycle assessment (LCA) study was conducted in 1969 by the Midwest Research Institute in the United States for the Coca-Cola company, for decision making on bottles packaging [10]. In the 1990s, much effort was placed in developing less time- and cost-intensive versions of the standard and simplified LCA method. In 1993, the first guidelines were published on LCA [11]. The issues addressed and approaches developed in order to reduce the labor and data intensity of the LCA method [12] is to have little cohesion seemingly remains in this type of assessment. Since then, LCA standards, methodologies, and applications have continued to be developed.

Key variables influence the LCA method, and two levels of incomplete LCAs exist, namely [13]

- Matrix mapping or screening LCAs – can be qualitative or semi-quantitative.

- Streamlined LCAs – are fully quantitative and are therefore preferentially applied to products or technologies with a relatively high readiness. Streamlining approaches usually involve reduced levels of detail or data quality [14–15]. These methods therefore focus on the most visibly apparent aspects of pollution such as packaging and landfilling [16].

In plastics waste management LCA is

- A powerful tool for assessing the overall environmental impact of practices.
- One of the most important analytical tools.
- Is sustained during both ecological design and life cycle management.
- Is a management technique.

9.2 Life Cycle Assessment – A Tool to Assess Waste

Reducing landfilling in favor of increasing energy and material recovery reduces environmental impacts and saves resources [17–19]. LCA has become a well-known and commonly used method. It is expected not only to lead to a comprehensive evaluation of the plastics waste under examination, but also to provide quantified results for comparison with other waste management systems or subsystems [20]. It finds extended application in waste management systems or subsystems [21–23].

Life cycle assessment (LCA) is a tool to assess all environmental exchanges from resources, energy, emissions, depositions, conversions, and waste. It is commonly used to quantify the environmental impact of different waste and also assesses environmental aspects and potential impacts associated with products or services. Furthermore, LCA methodology assesses the impact that several waste management operations have on the environment. In plastics waste, these operations involve sorting, recycling and incineration. Life cycle assessment is a powerful tool to address the prospects of plastics waste management and show the environmental impacts of generic activity.

LCA considers the entire life cycle of products or services from raw material acquisition through production, use and disposal. The life cycle analysis (LCA) of plastics waste management systems and their economic impacts generally report that recycling has the lowest environmental impact on global warming potential (GWP) and total energy use (TEU). The plastic materials have environmental and economic advantages over conventional materials throughout their life cycle with or without considering end-of-life (Figure 9.1) [24].

Developed countries can inspire developing and underdeveloped countries to accelerate the modernization of plastics waste management. It should be applied in waste management to ensure that plastics waste have effects which are least harmful to the environment.

- Prevention from production stage
- Preparing material for reuse
- Recycling by either mechanical or feedstock recycling
- Other recoveries such as energy recovery
- Disposal in terms of existing end-of-life treatments by reprocessing to new products (mechanical recycling).

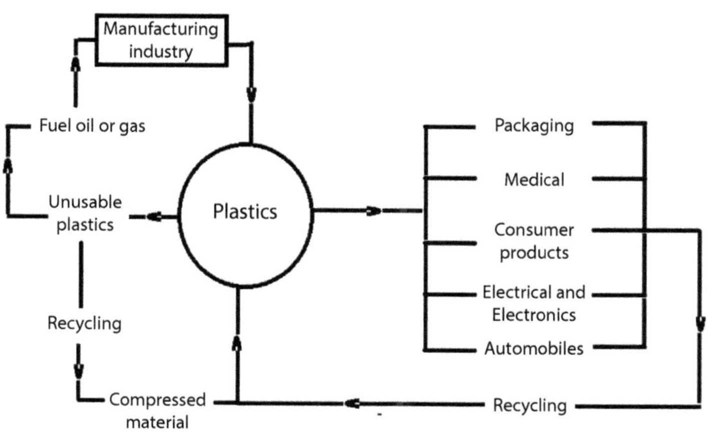

Figure 9.1 Schematic diagram of the life cycle of plastic materials.

9.3 Scientific Engineering

Life cycle assessment (LCA) provides the scientific background for engineering solutions. It has been used to highlight and quantify the environmental efficiency of innovative waste to production chain. The life cycle of a product or service includes raw material acquisition, production, use and end-of-life, including recycling and final disposal.

LCA can be used

- To improve the environmental performance of products/services
- To indicate relevant environmental monitoring and evaluation
- To compare alternative management options
- To evaluate the environmental profile of the product/system
- To understand the environmental significance of each aforementioned stage
- To provide various forms of assistance.

9.4 Purpose

Life cycle assessment has been used to highlight and quantify the eco-efficiency of innovative waste-to-production chain, with the objective of identifying solutions to be used in increasing energy efficiency. The purpose of life cycle assessment is

- Quantifying impacts on the environment and energy savings
- Improving eco-efficiency
- Increasing the credibility of recycling, reuse or recovery.

According to LCA, the environmental benefits quantified in waste are often directly proportional to the amounts of waste materials routed to a particular recycling technology; and the specific composition, properties, and presence of targeted resources in the waste materials.

9.5 Harmonization of LCA Method

Waste management is characterized by activities representing impacts to the environment, while the environmental benefits from resource recovery activities generally occur downstream of the waste system itself through displacement of other products and services [25]. As such, in-depth knowledge of the involved waste conversion and treatment processes, the emissions and properties of the outputs from these processes, and the relevant technology configurations and associated performances are all required to appropriately model the waste management system. LCA offers a methodology that may address precisely this conflict, waste LCA struggle with finding the right approach to reflect resource quality and displacement, rather than merely relying on material quantities [25, 26].

LCA is framed to assess the environmental impacts or negative impacts of the resource inputs and outputs which occurred in a functioning product system [27]. The prevention of waste generation is a central focus that must be maintained for current and future generations [28]. LCA focuses on assessing environmental performance through negative environmental indicators.

As far as plastics waste is concerned, the harmonization of the LCA method requires further formulation of technical recommendations such as

- LCA framework needs a consequential perspective to facilitate capturing not only direct impacts but also indirect consequences.
- LCA is still very challenging for plastics waste, but it is a stepping stone to paving the road to methodological and technical development.
- More application of case studies is recommended as lessons learned when integrating LCA.

9.6 Methodology

Life cycle assessment (LCA) determines the successful plastics waste management from recycling which comes from implementation of technological process. The hindrance in implementation of recycling include technology, management, and commercial issues. The environmental concerns and policy issues hinder their reuse before cleaning due to contaminants, hazardous substances such as pesticides, medicinal stocked may be stick on it. From manufacture to final disposal, the life cycle becomes complex through processing activities, service life and intended applications. In the life cycle, the severe and return of material from discard and waste plastics are to be managed by energy or recycling to add value to the technology and environment.

In environmental assessment methodologies, such as LCA, it has been assumed that collecting and recycling end-of-life materials prevents the production and therefore the impacts of similar materials from raw inputs and therefore also conserves finite resources. The theory is that the impacts of recycling are more than not offset by displaced virgin production, reducing net impacts.

Life cycle assessment (LCA) is a standardized method that strives to determine all of the environmental burdens connected to a product or activity. However, capturing all of these environmental flows is very difficult [29,30]. Although LCA is a powerful tool, it is not without limitations. For instance, the comprehensive nature of an LCA makes it costly and time intensive to perform [31–34]. LCA avoids the problem of moving the consequences of a process change, i.e., an apparent improvement in one part of a life cycle can merely lead to further problems at another time or place in the life cycle [35].

9.7 LCA Initiation

Life cycle assessment seeks to quantify the potential environmental impacts of waste plastics systems based on the flow of physical energy and material. The Life Cycle Initiative strategy is to work on the harmonization of LCA through four stages, namely [36]

- Goal and scope,
- Inventory analysis,

- Impact assessment, and
- Interpretation.

LCA is a method of assessing products or services and has been proven to be a valuable tool to document the environmental considerations that need to be part of the decision-making process with regards to environmental sustainability [37]. Energy recovery options utilize plastic waste streams as fuel for energy production, exploiting the high calorific content of plastics. The life cycle assessment (LCA) governs how the successful recycling of waste plastics is managed, i.e., recycling management is driven by the technological process used to recycle waste plastics.

9.8 LCA in Plastics Waste

Methods employed in LCA should be extended to, and enlarged from, the priority of avoiding negative impacts, to intensively promoting benefits in a way that supports sustainable development [38, 39]. Figure 9.2 illustrates plastics waste management in terms of life cycle of polymeric products. The disposal phase in LCA is often neglected or indicated as kilogram waste; however, it has been established that the disposal phase should be included in LCA studies [40]. The risk of decisions shifting burdens from the production or use phase to the disposal phase because of data gaps can therefore be diminished.

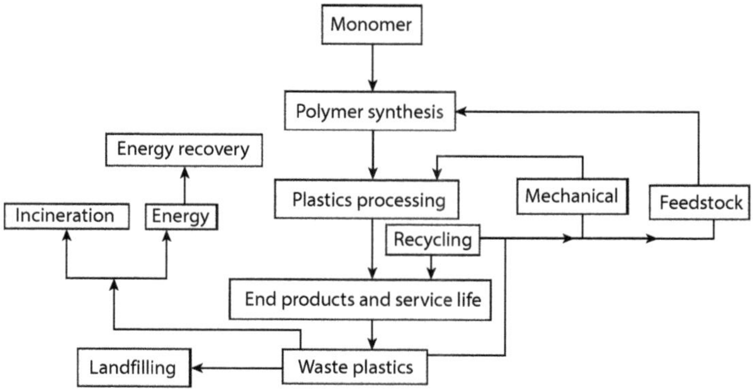

Figure 9.2 Plastics waste management life cycle.

LCA have been successfully utilized in the field of plastics waste management, for example, to assess differences in environmental performance between different waste incineration strategies [41] or related activities such as the flue gas cleaning process of plastic waste incinerators [36], to compare the environmental performance of different scenarios for the management of mixed plastics waste as well as that of specific plastic waste fractions [42–49].

9.9 Advantages of LCA

- LCA can be used to highlight and quantify the ecological efficiency of plastics waste-to-production chain.
- LCA can be used for comparisons of alternative options and can be integrated into a general decision-making process covering the environmental implications of a decision.
- LCA can be used to quantify the environmental impacts of different waste management options.

9.10 Shortcomings of LCA

- LCA is a data-extensive exercise.
- Inaccurate data, data gap and unrepresentative data are sources of uncertainty [50–53].
- LCA outputs are not sufficient to accurately assess the environmental impacts of materials that are reused.
- LCA of waste management systems is not a simple task since it involves a considerable amount of variables and extensive information.
- Standard LCA method struggles to evaluate technologies at low technological readiness levels and multitude of technology candidates at the experimental or laboratory scale due to insufficient data are available to perform an adequate LCA. The limitations make LCA unsuitable in the early stages of technology

development, as the necessary information is not usually available
- Data gaps are difficult to readily identify in typical LCA results unless explicitly highlighted by the practitioner. [54].

9.11 Environment Waste Auditing

Environment auditing is a new tool used by plastics processing industries in the management of plastics waste and environmental responsibilities. A common, practicable, and comparable measuring method to evaluate the efficiency of a recycle system is still missing. Effort is invested in optimizing the recycling system with a target value for efficiency, which includes carefully chosen recycling materials as well as proper processing equipment.

Environmental auditing has attracted attention as a tool to be used by industrial and government enterprises in waste plastics management and environmental responsibilities. However, this tool has neither a common understanding of the role of auditing nor the basic features of successful auditing programs. The objective of environmental waste auditing is to make life easier for both processors and consumers.

Input-output analysis (IOA) is an economic discipline that deals with the interdependence of plastics waste between and within industries and consumers through producing and consuming products and recycling. Since the uncertainty of life cycle initiative data is difficult to quantity, this method will not solve an underlying problem. The effort required for understanding and interpreting the data is based on focusing on the complexity of a subject.

LCA involves compiling an inventory of inputs and outputs of the relevant product system, which are then evaluated, along with interpretation of the results of the inventory analysis and impact assessment in relation to the objective of the study, to determine the potential impact of the product/service on the environment [55].

Technology development and design choices made early in the process have a significant effect on the overall environmental performance of the final product. The development phase that the results of an LCA would paradoxically have the greatest influence on the technology [56]. A systematic framework allowing for an efficient and transparent waste plastics of multiple technological candidates early in development would therefore be useful.

9.12 Waste Prevention

Waste prevention—that is, reuse, source reduction, and cleaner waste materials—is possibly the most important challenge of them all, both in the context of LCA and in society itself. Prevention effects are inherently difficult to quantify and regulate; in many countries, this has resulted in somewhat limited attention from regulatory agencies [57]. As waste LCA typically apply a zero-burden assumption, in which the upstream environmental impacts associated with production, distribution, and use phases of materials in the waste are excluded from the system boundaries, evaluating environmental benefits from waste prevention is not straightforward [58–61].

Plastic products need to be responsibly managed throughout their life cycle so that they are prevented from escaping into the environment by proper waste management, mindful product design and responsible behavior. A number of measures outlined below can be implemented on a wide scale, from the end-user to the manufacturers, to manage the design, use and recycling of plastic materials.

- Source reduction activities
- Replacement of single-use products with eco-friendly material
- Life extension with better design and reuse
- Management responsibility throughout the life cycle
- Avoiding unnecessary use where possible
- Reducing number of components of plastics
- Additives standardization to achieve easy removal of plastics from other chemicals
- Sustainable and effective separation, sorting and cleaning
- Government actions to achieve a broader engagement
- Research and development for new and sustainable methods of recycling

9.13 Remarks

Every new concept and tool has a user learning curve at the beginning. LCA is an advancing and changing field that not only demands knowledge, but also consistent, usable guidelines in order to continue to grow. Within a given scope in waste plastics, individual life cycle phases are important in regard to product or recycling or inventory. The assessment process regards the degree of detail which can be related to the entire waste plastics recycle system or to its products or to individual stages of recycling. LCA represents important steps on the way to improved insight into methodological choices such transparent results, robust and unbiased conclusion. Plastics waste LCA as an important component of the work of industrial ecology of when and how to close resource loops. However, LCA paves the road for high quality and relevant decision support for the benefit of society.

References

1. Toso, E.A.V. and Alem, D., Effective location models for sorting recyclables in public management. *Eur. J. Oper. Res.*, 234, 839–860, 2014.
2. Abeliotis, K., Kalogeropoulos, A., Lasaridi, K., Life cycle assessment of the MBT plant in Ano Liossia, Athens, Greece. *Waste Manag.*, 32, 1, 213–219, 2012.
3. Cadena, E., Colón, J., Artola, A., Sánchez, A., Font, X., Environmental impact of two aerobic composting technologies using life cycle assessment. *Int. J. Life Cycle Ass.*, 14, 5, 401–410, 2009.
4. Bergsdal, H., Strømman, A.H., Hertwich, E.G., Environmental assessment of two waste incineration strategies for central Norway. *Int. J. Life Cycle Ass.*, 10, 4, 263–272, 2005.
5. Tonini, D., Albizzati, P.F., Astrup, T.F., Environmental impacts of food waste: Learnings and challenges from a case study on UK. *Waste Manag.*, 76, 744–766, 2018.
6. Ripa, M., Fiorentino, G., Vacca, V., Ulgiati, S., The relevance of site-specific data in life cycle assessment (LCA). The case of the municipal solid waste management in the metropolitan city of Naples (Italy). *J. Clean. Prod.*, 142, 445–460, 2017.
7. Finnveden, G., Johansson, J., Lind, P., Moberg, Å., Life cycle assessment of energy from solid waste—Part 1: General methodology and results. *J. Clean. Prod.*, 13, 213–229, 2005.

8. Lea, W.R., Plastic incineration versus recycling: A comparison of energy and landfill cost savings. *J. Hazard. Mater.*, 47, 295–302, 1996.
9. Eriksson, O. and Finnveden, G., Plastic waste as a fuel - CO2-neutral or not? *Energy Environ. Sci.*, 2, 907–914, 2009.
10. Bauman, H. and Tillman, A.-M., *The Hitch Hiker's Guide to LCA: An Orientation in Life Cycle Assessment Methodology and Applications*, 1st ed., Studentlitteratur AB, Lund, Sweden, 2004.
11. Consoli, F., Allen, D., Boustead, I., Fava, J., Franklin, W., Jensen, A.A., de Oude, N., Parrish, R., Perriman, R., Postlethwaite, D. *et al.*, *Guidelines for Life-Cycle Assessment: A Code of Practice*, 1st ed., F. Consoli (Ed.), Society of Environmental Toxicology and Chemistry (SETAC), Pensacola, FL, USA, 1993.
12. Curran, M.A. and Young, S., Report from the EPA conference on streamlining LCA. *Int. J. Life Cycle Ass.*, 1, 1, 57–60, 1996.
13. Wenzel, H., Application dependency of LCA methodology: Key variables and their mode of influencing the method. *Int. J. Life Cycle Ass.*, 13, 5, 281–288, 1998.
14. Moberg, A., Borggren, C., Ambell, C., Finnveden, G., Guldbrandsson, F., Bondesson, A., Malmodin, J., Bergmark, P., Simplifying a life cycle assessment of a mobile phone. *Int. J. Life Cycle Ass.*, 19, 5, 979–993, 2014.
15. Arzoumanidis, I., Salomone, R., Petti, L., Mondello, G., Raggi, A., Is there a simplified LCA tool suitable for the agri-food industry? An assessment of selected tools. *J. Clean. Prod.*, 149, 406–425, 2017.
16. Guinée, J.B., Heijungs, R., Huppes, G., Zamagni, A., Masoni, P., Buonamici, R., Ekvall, T., Rydberg, T., Life cycle assessment: Past, present, and future. *Environ. Sci. Technol.*, 45, 1, 90–96, 2011.
17. Rigamonti, L., Grosso, M., Giugliano, M., Life cycle assessment for optimising the level of separated collection in integrated MSW management systems. *Waste Manag.*, 29, 934–944, 2009.
18. Cherubini, F., Bargigli, S., Ulgiati, S., Life cycle assessment LCA of waste management strategies: Landfilling, sorting plant and incineration. *Energy*, 34, 2116–2123, 2009.
19. Giugliano, M., Cernuschi, S., Grosso, M., Rigamonti, L., Material and energy recovery in integrated waste management systems. An evaluation based on life cycle assessment. *Waste Manag.*, 31, 2090–2101, 2011.
20. ISO 14040:2006 (E), Environmental management – Life cycle assessment – Principles and framework; International Standards: ISO 14044:2006 (E) (2006) Environmental management – Life cycle assessment – requirements and guidelines, 2006.
21. Martínez-Blanco, J., Colón, J., Gabarrell, X., Font, X., Sánchez, A., Artola, A., Rieradevall, J., The use of life cycle assessment for the

comparison of biowaste composting at home and full scale. *Waste Manag.*, 30, 983–994, 2010.

22. Koroneos, C.J. and Nanaki Evanthia, A., Integrated solid waste management and energy production – A life cycle assessment approach: The case study of the city of Thessaloniki. *J. Clean. Prod.*, 27, 141–150, 2012.
23. Wang, L., Templer, R., Murphy, R.J., A Life Cycle Assessment (LCA) comparison of three management options for waste papers: Bioethanol production, recycling and incineration with energy recovery. *Bioresour. Technol.*, 120, 89–98, 2012.
24. Bernardo, C.A., Simões, C.L., Lígia, M., Pinto, C., Environmental and economic life cycle analysis of plastic waste management options. A review. *AIP Conference Proceedings*, 1779, 140001, 2016.
25. Vadenbo, C., Hellweg, S., Astrup, T.F., Let's be clear(er) about substitution: A reporting framework to account for product displacement in life cycle assessment. *J. Ind. Ecol.*, 21, 5, 1078–1089, 2017.
26. Zink, T., Geyer, R., Startz, R., A market-based framework for quantifying displaced production from recycling or reuse. *J. Ind. Ecol.*, 20, 4, 719–729, 2016.
27. Clift, P.R., *Social life cycle assessment: What are we trying to do? In Pre-Proceedings of the 4th International Seminar on Social LCA*, pp. 11–16, CIRAD, Montpellier, France, 2014.
28. Schaubroeck, T., A Revision of What Life Cycle Sustainability Assessment Should Entail Towards Modeling the Net Impact on Human Well-Being. *J. Ind. Ecol.*, 21, 1–14, 2017.
29. Rebitzera, G., Ekvallb, T., Frischknechtc, R., Hunkelerd, D., Norrise, G., Rydbergf, T., Schmidtg, W.-P., Suhh, S., Weidemai, B.P., Pennington, D.W., Life cycle assessment Part 1: Framework, goal and scope definition, inventory analysis, and applications, *Environment International*, 30, 701–720, 2004.
30. Stefanie Hellweg and Llorenç Milà i Canals, Emerging approaches, challenges and opportunities in life cycle assessment, *Science* 344, 1109, 2014.
31. Gradel, T.E., Benkovitz, C.M., Keene, W.C., Lee, D.S., Marland, G., Global emissions inventories of acid-related compounds, *Water, Air and Soil Pollution*, 85, 25–36, 1995.
32. Keith A. Weitz, joel Ann Todd, Mary Ann Curran, Melissa J. Malkin, Streamlining Life Cycle Assessment: Considerations and a Report on the State of Practice, *Int. J. LCA*, 1(2), 79–85, 1996.
33. Elisabeth Hochschorner 1. and G6ran Finnveden, Evaluation of Two Simplified Life Cycle Assessment Methods, *Int J LCA*, 8(3), 119–128, 2003.

34. Gδran Finnveden, Michael Z. Hauschild, Tomas Ekvall, Jeroen Guinée d, Reinout Heijungs, Stefanie Hellweg, Annette Koehler, David Pennington, Sangwon Suh, Recent developments in Life Cycle Assessment, *Journal of Environmental Management*, 91, 1–21, 2009.
35. Use of Life Cycle Assessment (LCA) as a Policy Tool in the Field of Sustainable Packaging Waste Management, *The European Organization for Packaging and the Environment*, Etterbeek, Belgium, 1999.
36. Rizal, T.F., Patrick, L., Luca, S., Mohammad, D.H., Ben, A., Exploring the Current Challenges and Opportunities of Life Cycle Sustainability Assessment. *Sustainability*, 11, 636–652, 2019.
37. Gheewala, S.H., Application of Life Cycle Assessment to. Cleaner Production. *Int. Energy J.*, 4, 1, 5, 2003.
38. Sala, S., Farioli, F., Zamagni, A., Progress in sustainability science: Lessons learnt from current methodologies for sustainability assessment: Part 1. *Int. J. Life Cycle Assess.*, 18, 1653–1672, 2013.
39. Zamagni, A., Pesonen, H., Swarr, T., From LCA to Life Cycle Sustainability Assessment: Concept, practice and future directions. *Int. J. Life Cycle Assess.*, 18, 1637–1641, 2013.
40. Doka, G. and Hischier, R., Waste Treatment and Assessment of Long-Term Emissions. *Int. J. Life Cycle Ass.*, 10, 1, 77, 2004.
41. Bergsdal, H., Strømman, A.H., Hertwich, E.G., Environmental Assessment of Two Waste Incineration Strategies for Central Norway. *Int. J. Life Cycle Ass.*, 10, 4, 263, 2005.
42. Chevalier, J., Rousseaux, P., Benoit, V., Benadda, B., Environmental assessment of flue gas cleaning processes of municipal solid waste incinerators by means of the life cycle assessment approach. *Chem. Eng. Sci.*, 58, 2053, 2003.
43. Denison, R., Environmental life-cycle comparisons of recycling, landfilling, and incineration: A review of recent studies. *Ann. Rev. Energy Environ.*, 21, 191, 1996.
44. Finnveden, G. and Ekvall, T., Life-cycle assessment as a decision-support tool—The case of recycling versus incineration of paper. *Resour. Conserv. Recy.*, 24, 235, 1998.
45. Finnveden, G., Johansson, J., Lind, P., Moberg, Å., *Life cycle assessments of energy from solid waste. fms 137*, FOA-B-00-00622-222—SE. Report from Environmental Strategies Research Group, Stockholm, Sweden, 2000.
46. Mendes, R.M., Aramaki, T., Hanaki, K., Comparison of the environmental impact of incineration and landfilling in São Paulo City as determined by LCA. *Resour. Conserv. Recy.*, 41, 47, 2004.

47. Arena, U., Mastellone, M.L., Perugini, F., The environmental performance of alternative solid waste management options: A life cycle assessment study. *Chem. Eng. J.*, 96, 207, 2003.
48. Ross, S. and Evans, D., The environmental effect of reusing and recycling a plastic-based packaging system. *J. Clean. Prod.*, 11, 561, 2003.
49. Sonesson, U., Björklund, A., Carlsson, M., Dalemo, M., Environmental and economic analysis of management systems for biodegradable waste. *Resour. Conserv. Recy.*, 28, 29, 2000.
50. Kaplan, O., Barlaz, M., Ranjithan, R., A procedure for life-cycle-based solid waste management with consideration of uncertainty. *J. Ind. Ecol.*, 8, 4, 155–72, 2005.
51. De Feo, G. and Malvano, C., The use of LCA in selecting the best MSW management system. *Waste Manag.*, 29, 6, 1901–15, 2009.
52. United Nations Environmental Programme, Waste and Climate Change: Global trends and strategy framework, Division of Technology, Industry and Economics International Environmental Technology Centre, pp. 6–34, Shiga, Osaka, Japan, 2010.
53. Bojaca, C.R. and Schrevens, E., Parameter uncertainty in LCA: Stochastic sampling under correlation. *Int. J. Life Cycle Ass.*, 15, 3, 238–46, 2010.
54. Hetherington, A.C., Borrion, A.L., Griffiths, O.G. and McManus, M.C., Use of LCA as a development tool within early research: challenges and issues across different sectors, *The International Journal of Life Cycle Assessment*, 19, 1, 130–143, 2014.
55. ISO 14040: 1997(E), *Environmental Management - Life Cycle Assessment - Principal and Framework*, International Organization for Standardization, Geneva, Switzerland, 1997.
56. Hellweg, S. and Mila i Canals, L., Emerging approaches, challenges and opportunities in life cycle assessment. *Science*, 344, 6188, 1109–1113, 2014.
57. Fischer, S., *Stepping up waste prevention - challenges and opportunities for National Waste Prevention Progreammes*, Policy Brief No. 7, pp. 3–15, 2017.
58. Dormer, A., Finn, D.P., Ward, P., Cullen, J., Carbon footprint analysis in plastics manufacturing. *J. Clean. Prod.*, 51, 133–141, 2013.
59. Dolci, G., Nessi, S., Rigamonti, L., Grosso, M., Life cycle assessment of waste prevention in the delivery of pasta, breakfast cereals, and rice. *Integr. Environ. Assess. Manag.*, 12, 3, 445–458, 2016.
60. Nessi, S., Rigamonti, L., Grosso, M., LCA of waste prevention activities: A case study for drinking water in Italy. *J. Environ. Manage.*, 108, 73–83, 2012.
61. Gentil, E.C., Gallo, D., Christensen, T.H., Environmental evaluation of municipal waste prevention. *Waste Manag.*, 31, 12, 2371–2379, 2011.

10 Case Studies

A case study begins with a description of the background of the problem and the situation is then analyzed. In a manufacturing plant, a problem regarding quality requires changing various parameters and mapping input to output data as no accurate theoretical model for solving interrelated process problems exists; input and output problems occur due to machine and operator performance.

Case study methodology is well suited for many types and aspects of plastics waste management. Research into plastics waste is a contemporary development and is complicated by the fact that waste components/processes are difficult to study in isolation. Therefore, case studies do not generate the same results on recycling as controlled experiments; however, they provide a deeper understanding of the phenomena under study.

10.1 Waste Dump and Health Hazards

Furthermore, the collection and composition of packaging waste gathered by impoverished people, in exchange for money, varies and segregation adds to the cost of recycling. Waste dumping sites create a nuisance to the public and harbor disease due to the lack of landfilling sites. In addition, waste dumping sites attract animals which cause further health hazards. The majority of cities in India have landfill sites within the city limit, as elevated transportation prices and roads with potholes mean it costs more to carry the waste beyond the city limit; hence the disease rates are unfortunately increasing on a daily basis.

Muralisrinivasan Natamai Subramanian (ed.) Plastics Waste Management, 2nd Edition (199–218) © 2019 Scrivener Publishing LLC

10.2 Utilization of Plastics Waste

With a holistic approach to waste management is needed by collecting the segregated waste at source, and characterizing and processing the waste in order to produce value-added products. Household waste contributes a major portion to the total MSW; in addition, waste from commercial establishments, hospitals and hotels, sweepings from the street and construction activities add to MSW. Creating awareness through the participation of local people in programme- and community-based waste management systems would be a useful tool in the management of plastics waste. Monitoring the generation of plastics waste and characterizing and assessing the physical and chemical characteristics of the waste yield essential information.

The high export share of the plastics industry in Europe along with the United States and Japan, excluding internal business, is seen from a different angle. Over the last decades there has been strong growth of plastics processing in Europe, the United States and Japan, generating about 55 million tons of post-consumer plastic waste [1].

Municipal solid waste contains postconsumer plastics which consist of mainly high-density polyethylene (HDPE), low-density polyethylene (LDPE), polypropylene (PP), polystyrene (PS), polyvinyl chloride (PVC), and polyethylene terephthalate (PET) [2]. Post-consumer plastic waste generation in Europe is approximately 25 million tonnes per year, of which 6.6 million tonnes are recycled mainly through mechanical recycling, 8.9 million tonnes are processed by energy recovery facilities, and 9.6 million tonnes are disposed of in landfills [3]. MSW can aggravate different types of environmental pollution such as air, water, including groundwater, and soil pollution, which affects public health as the waste forms a breeding ground for various disease-carrying vectors such as flies, mosquitoes, rats and others, and indicates inappropriate and unscientific management [4]. These waste products were previously dumped in landfill sites, a non-sustainable and environmentally questionable option. The number of landfill sites and their capacity are moreover decreasing rapidly and in most countries the legislation on landfills is becoming increasingly stringent.

10.2.1 Europe

The sustainability of municipal solid waste (MSW) management has been predominantly driven in Europe. The European waste and natural resources policy is governed by the need to protect human health and the environment, and aims to reduce any negative impact on the environment via economic, technological and social constraints.

Waste generally means any substance or object which the holder discards, intends to or is required to discard, and waste management means the collection, transport, recovery and disposal of waste, and includes the supervision of such operations and aftercare of disposal sites [5].

The European Council encourages the prevention or reduction of waste production as the preferred solution. In 1994, acceptable levels of plastics waste recycling and recovery were established by the European Union (EU) regarding packaging and packaging waste; for all packaging waste streams, including plastic, a minimum recycling target of 15 wt% was set [6].

Later it was amended to 22.5 wt% for plastics, which included material that was recycled back into plastic with no distinction among the various streams of plastic waste, such as industrial, commercial, office, service or household [7].

The European Commission's 'Thematic Strategy' on the prevention and recycling of waste established new aims and objectives for EU waste policy. In addition, it put forward a vision of an EU 'recycling society' as its long-term goal [8].

10.2.2 India

In India, there are numerous deficiencies in many of the current practices of plastics waste handling. During the rainy season, waste management cannot be carried out properly which impacts public health. This situation is mainly due to limited human and financial resources, and the fact that machinery cannot effectively carry out various activities; to overcome this problem effective strategies and guidelines must be established.

Most of the manufacturing of plastics and additives raw materials occurs in a facility located in Gujarat, India; with facilities for medium- and small-scale plastics processing being located in different states. However, economic development founded upon manufacturing and industrialization has resulted in a high degree of environmental pollution. This is particularly true for Gujarat, not only as a result of its having become a manufacturing hub, but also the rapid industrial transformation achieved at the expense of the environment and society. The state and corporations focused mainly on industrial development through massive manufacturing and ignored environmental regulations. Consequently, the district of Bharuch in the Gujarat state is a highly polluted zone in India [9]. There is a dangerous situation of contamination of plastics particularly in India due to mixing of plastics waste in tar for road laying. Due to weathering, these mixed plastics become microplastic particles and spread throughout the region. This mixing will be become a big issue in India in the near future due to the permanent pollution caused by the inseparable portion of plastics retained in bitumen. Waste disposal in the case of plastics waste with tar laying acknowledge its risk, but portray it as a source of economic opportunity. In India, environmental accounting is relatively low at the industrial level [10]. In some of the industries, environmental management is only provided at minimum levels.

10.2.3 Japan

Japan is still entertaining the world with sudden moves. A stagnant GDP, rising unemployment, and ever higher budget deficits resulting from failed attempts at overcoming economic stagnation and contributing to a frighteningly high debt-to-GDP ratio have become hallmarks of Japan. In Japan, area availability is extremely limited. The principal products present in plastics waste are polyethylene, polyvinylchloride, polypropylene and polystyrene. Among these plastics materials, the combustion of polyvinylchloride (PVC) matter generates dioxin and its associated problems. Out of total plastics waste, these plastics account for approximately 70 percent:

- 38 percent in reuse
- 15 percent for power generation
- 12 percent as heat source – thermal energy recycling
- 11 percent recycling

Of the total amount of solid waste, 61 percent is disposed of in landfills and by combustion.

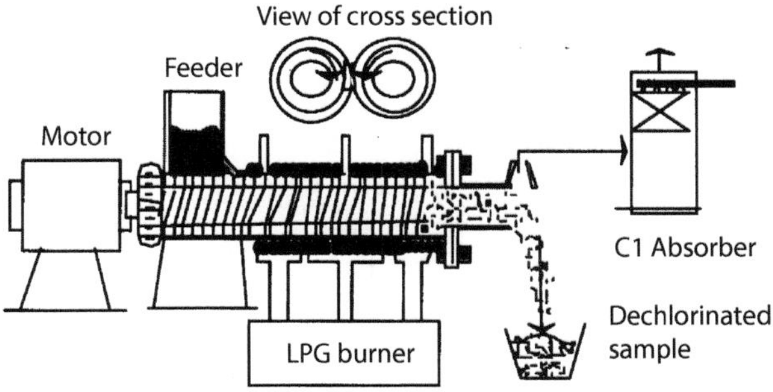

Figure 10.1 First stage of chlorine removal from waste plastics with twin-screw extruder. [Reproduced with permission from Kiyoshi Saito, Hideo Narita, *J Mater Cycles Waste Manag*, 2001, 3, 93–98 [11]. Copyright © 2001 Springer-Verlag. All rights reserved].

Recycling of waste plastics with polyvinylchloride (PVC) has been established for the production of solid and liquid fuel in Japan. The process includes a process to remove chlorine from the waste plastics, which is a two-stage catalytic pyrolysis process. In the first stage, the dechlorination equipment consists of a twin screw extruder with a heating element as hydrogen chloride (Figure 10.1).

After dechlorination, it could be used as solid fuel with high calorific value due to no formation of toxic gas like dioxin and other reactions, which reduces the calorific value. The solid fuel product could be used in oil production process. The light oil fraction has a boiling point between that of kerosene and gasoline (Figure 10.2).

Selection of suitable catalyst for cracking reactions is a difficult problem due to limitations in terms of operating conditions, feedstock properties, catalyst performance, and desired products. The suitability of solid heterogeneous catalysts is based on the essential characteristics of activity, selectivity, ease of regeneration, and strength. The technology with natural and synthetic zeolite as catalyst makes it possible to produce a non-polluting, high-calorific solid fuel and a liquid fuel very efficiently [11].

10.2.4 France

In 2016, France banned plastic plates, cups, and utensils with a law that will go into effect in 2020, becoming the first country in the world to do so. Exceptions will be allowed for items made of

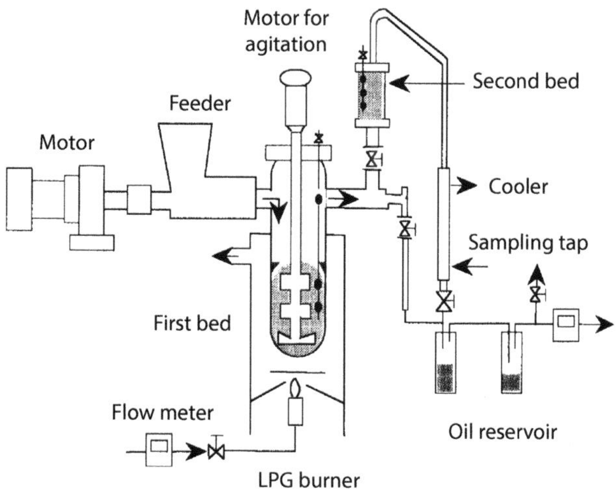

Figure 10.2 Schematic diagram of two-stage catalytic processing. [Reproduced with permission from Kiyoshi Saito, Hideo Narita, *J Mater Cycles Waste Manag*, 2001, 3, 93–98 [11]. Copyright © 2001 Springer-Verlag. All rights reserved].

compostable, bio-sourced materials [12]. There will be a net benefit of discarding less single-use plastic materials; however, there are serious concerns about their replacements. Recycling plastic materials seems eco-friendlier than replacement of plastic with paper, bioplastic or, in particular, cotton.

10.2.5 Other Countries

In 2015, the United Kingdom stopped handing out plastic bags for free [13]. This resulted in almost a six billion reduction in the number of plastic bags handed out in six months [14].

Saudi Arabia is one of the major producers of plastic in the world with total production capacity of around six million metric tons per year. It is one of the largest oil-producing countries in the world. Since petroleum residue is hydrogen deficient, co-processing with waste plastics would be another way to recycle plastics waste into useful products [15, 16].

In Indonesia, households separate their waste and bring it to deposit into individual accounts at a local waste bank. A waste bank operates in terms of selling aggregated plastics waste to depositors to selling agents, and return part of the sale proceeds to agents in the form of cash, in-kind goods, or other services [17].

Considering that solid waste management in Vietnamese cities is challenged by rapid economic growth, as well as growth of urban population, alternatives to landfilling are needed to cope with limited availability of land, limited economic resources and the necessity of lowering greenhouse gas emissions.

10.3 Use of Case Studies

Cases and case studies can be powerful aids. Both can be used to train professionals in fields such as education, public administration, management, medicine, social work and caring occupations. Case studies can be an engaging way to evaluate the feasibility of possible solutions [18].

Furthermore, existing work on the empirical research methodology of plastics waste management has a strong focus on experimental research. Statements, ideas, points-of-view and research decisions are surprising, perhaps outrageous; however, they are challenging and enlightening. All research has a tendency towards quantitative approaches, hence, the literature involving case studies of waste plastics and qualitative methods is limited.

The type of learning that occurs when immersed in situations and experiences varies from situation to situation. Case studies are commonly used in the fields of psychology, sociology, political science, social work, business and community planning, [19]. To increase our knowledge regarding plastics waste management, case studies are conducted for social, community and related phenomena, which allows comparison with the research objectives in plastics waste management.

The use of environmental case studies, related to plastics waste, enables analysis of the extent to which contamination may adversely impact property value through the effect previously referred to and defined as 'environmental stigma' [20].

Case studies, as a tool for generating and testing theories, have provided the general management field with some of its most groundbreaking insights [21, 22].

Therefore, a case study is:

- Not simply a data-gathering technique.
- A method which incorporates a number of strategies.

- Incorporated with sources of evidence [23] including documents such as agendas, meeting minutes, reports, diaries with archival records, calendars, service records, direct and participant observation, and interviews which may be focused or open ended [24–26].
- The triangulation of data that results from the use of multiple methods and data sources to examine one phenomenon [27].
- Serving different purposes which may be exploratory, explanatory or descriptive [28, 29].

10.4 Property Value

With regards to plastics waste, there can be three effects on property value that may result from environmental contamination:

- Cost: a lowering of house price due to the costs to remediate a contaminated property to appropriate regulatory standards.
- Use: limitations on the use of properties that may be impacted by environmental contamination.
- Adverse effects on value due to increased perceptions of environmental risk by relevant market participants.

The most difficult to analyze and quantify, due to a lack of supporting evidence, is the effect on the property market. Risk or adverse effects can be quantified with clear and convincing market data [30].

10.5 Case Study 1: Plastics Waste from the Electric and Electronic Field

10.5.1 Concept

Typically, the plastics waste created by the electric and electronic field is 15–30%, hence, the recycling quotas cannot be fulfilled solely by state-of-the-art metal and glass recycling. The European 'Waste Electric and Electronic Equipment (WEEE) Directive' defines strict recycling and recovery quotas (EC2003a) namely, 70–80% for recovery and 50–75% for recycling strategies, which will eventually apply to all countries and will increase in the future; therefore, it is imperative to provide and develop economic recovery systems for plastics. In Europe, more than 6 million tonnes of WEEE is produced annually, which contains more than 1 million tonnes of plastic [31].

10.5.2 Objective

The separation of solvents from specific plastics, e.g., polyvinyl chloride (PVC) is separated from other polymers such as polyethylene (PE), polyoxymethylene, polycarbonate (PC) and so on, within electrical and electronic waste materials is achieved using its density and solubility via the cyclohexanone process. The following points need to be assessed when using a solvent process method:

1. The economic benefit of different material recycling approaches.
2. The presence of additives such as brominated flame retardants (BFR), including brominated biphenyls and polybrominated diphenyl ether [32].
3. Determination of polymer properties such as the density and solubility.

10.5.3 Methodology

The first stage of the cyclohexanone or CreaSolv® process [Figure 10.3] involves density separation, which allows the separation of PVC or styrenics, respectively, from other identified polymers in waste originating from electrical and electronic equipment. The density of PVC or styrenics overlaps with filled polypropylene (PP) and rigid polyurethane, and PVC and styrenics even overlap each other; therefore, density-based polymer separation alone will not permit pure recycled polymers to be obtained.

The majority of electrical and electronics waste contains PVC as it is used in electrical and electronics wiring as a sheathing material and insulator, and in electrical fittings along with other materials such as PE, polyoxymethylene, PC and so on, which form part of the waste material.

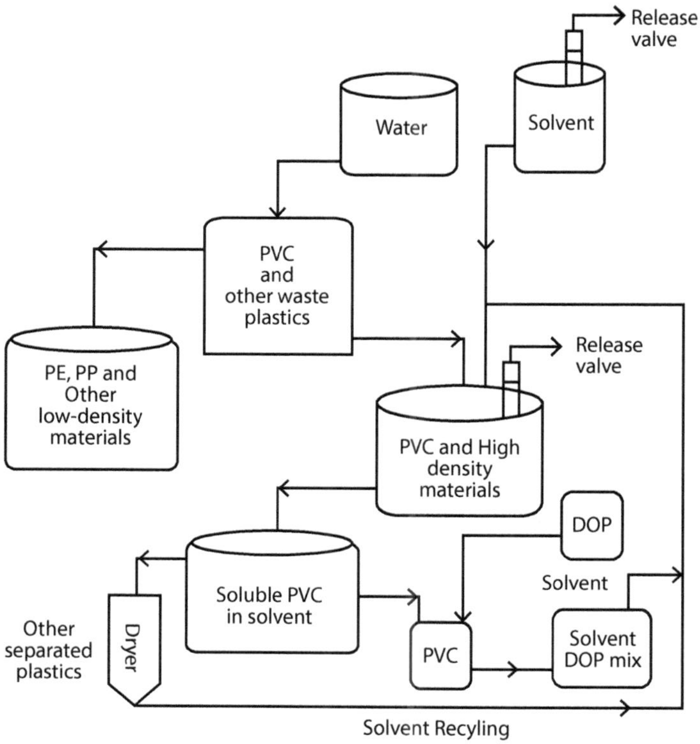

Figure 10.3 Separation of PVC from other plastics using the solvent technique, DOP (Dioctylphthalate).

10.5.4 Experimental Method

Figure 10.3 illustrates a flow diagram of the density and solvent separation of PVC from other plastic materials. During density separation, there is the possibility that materials of similar density may separate and settle along with the PVC. Using this method 5–8% of the PVC is soluble, and therefore a 12 kg sample can be used for the separation of PVC, which is a major quantity in comparison with other plastics.

After the separation of PVC from the mixture of waste originating from electrical and electronic equipment, the occlusion of plasticisers needs to be addressed.

If the above process is used, as opposed to a similar method using shredder residues produced during processing, WEEE is excluded from recycled material due to the variety of polymeric materials and presence of BFR; in addition, WEEE might also contain banned polybrominated diphenyl ethers or toxic polybrominated dioxins and furans [31]. The separation of residual non-styrenic polymers can be achieved using the CreaSolv® process, a solvent-based technology. In addition, this process allows the removal of non-polymeric materials, such as dust, metals and glass splinters, as well as unwanted additives and contaminants, such as BFR, polybrominated dibenzo-p-dioxins and polybrominated dibenzofurans, from the polymer solution [33]. Both density and solubility methods enable the separation of PVC from other materials [31]. Figure 10.4 illustrates the two-stage density separation of a medium density styrenics fraction followed by the extractive CreaSolv® process; this method uses a 11.5 kg sample.

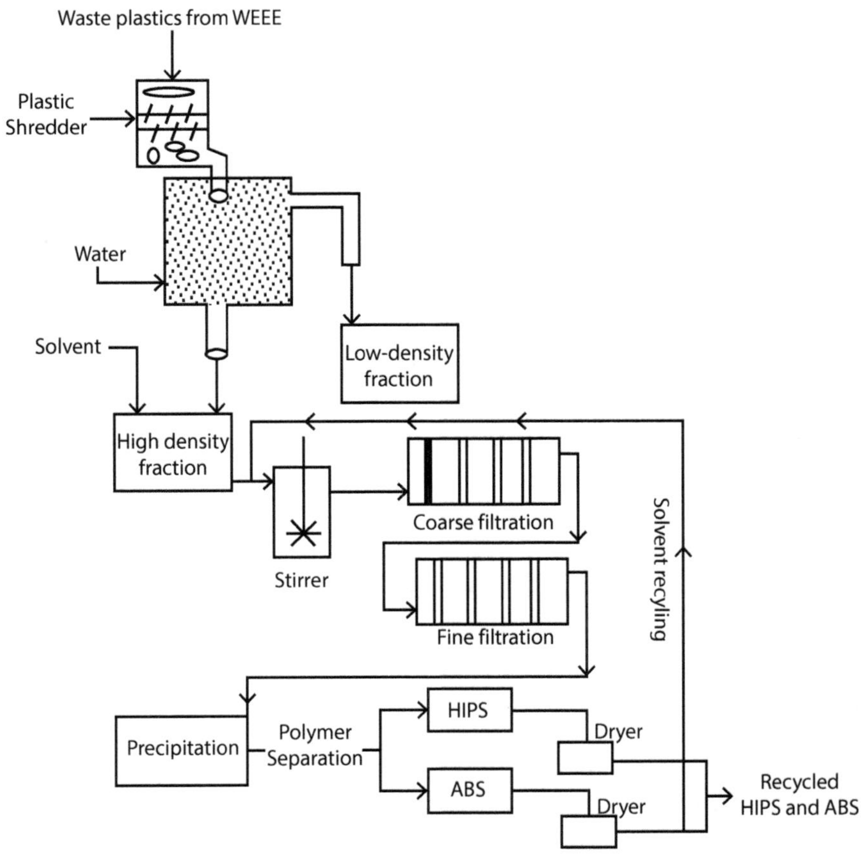

Figure 10.4 Illustration of density separation of WEEE waste, ABS (Acrylonitrile-butadiene-styrene) and HIPS (High-impact polystyrene).

10.5.5 Results

In both methods, the separated plastic materials are not in pure form. ABS and HIPS are present in the mixture and the two cannot be separated from each other. This styrenics fraction can be recycled either as it is or added to virgin material. With regards to PVC, DOP is present but it is not possible to separate the two; therefore, PVC can be used as a recycled material in flexible applications.

The solvent separation technique is quite costly and the separation of even small quantities of waste material requires high levels of solvents. As the solvent contamination increases, the solvation power of the recycled solvent decreases and become less effective, until it finally becomes a pollutant and requires special disposal as it contains toxic materials eluted from the waste plastics.

10.5.6 Conclusion

The plastic material separated from WEEE via density separation is used in blends along with other polymeric materials which are compatible with each other. Even for the separation of styrenics or PVC, the technological cost is higher using the solvent technique. This case study reveals that there may be a possibility for the separated plastics to be a valuable raw material for polymer recycling. The intermediate fraction from low-cost density separation serves in both methods as an appropriate input for the extraction of specific plastics from mixed plastics. Furthermore, there is also the requirement to prove the cost effectiveness on a larger scale using different separation techniques.

10.6 Case Study 2: Plastics Waste from the Automobile Industry

10.6.1 Background

Today, plastic used in the automobile industry is receiving some attention from academics; however, the majority of the focus on waste plastics is by the automobile company along with the utilization of waste plastic in its products. The automobile industry has improved production using process development; one step involved the utilization of plastics waste to create economic models for the automobile market. Economy can be realized by implementing the recycling of waste as opposed to its disposal.

10.6.2 Design

Design attempts to combine the elements of art and science. The first consideration in any design and optimization problem is to decide the boundaries of the system. Optimized process design in plastics waste recycling considers the costs of manufacturing, processing technique and the minimization of waste.

10.6.3 Disposal and Recovery

10.6.3.1 Recycling of Bumpers

The plastics waste from an automobile is around 10% of the total weight of the automobile; bumpers account for 10% of the total weight of plastic, making this an important target for recycling. The performance of bumper recycling depends on the complete removal of paint. Recycling will become a prerequisite in products such as automotive components. Crushing and pulverizing results in limited noise and dust; in addition, pulverizing is comparatively easy and enables recycled products to maintain certain properties including strength.

The evaluation of automotive materials involves recovering parts from cars which have been used for a long period and determining the properties of the plastics waste. Figures 10.5 and 10.6 illustrate the disposal of polymer parts which are present in automobiles and the recycling of bumpers, respectively.

The parts obtained from scrap automobiles are sorted through for plastic parts. The appearance and weight of the parts are measured and then subjected to crushing and pelletizing. Test specimens are prepared by molding or cutting the parts and the test pieces are evaluated.

10.6.4 Inference

ABS is widely used for the interior and exterior parts of automobiles, and parts such as the console box, glove box, radiator grill, door

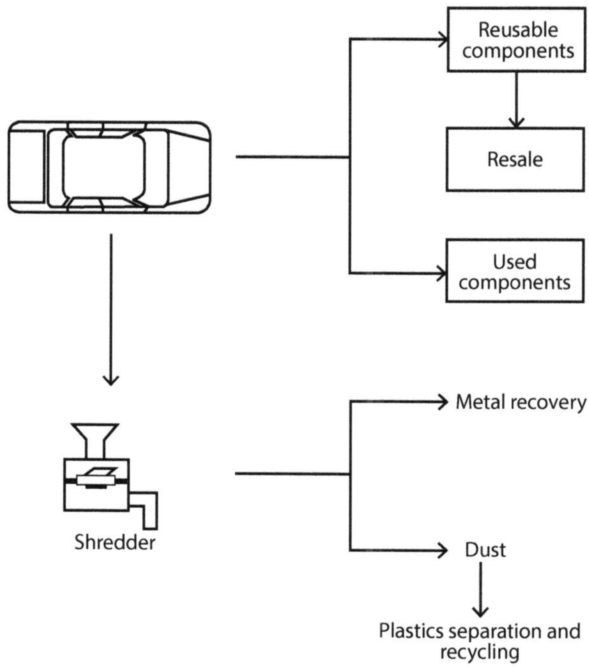

Figure 10.5 Disposal of automobiles.

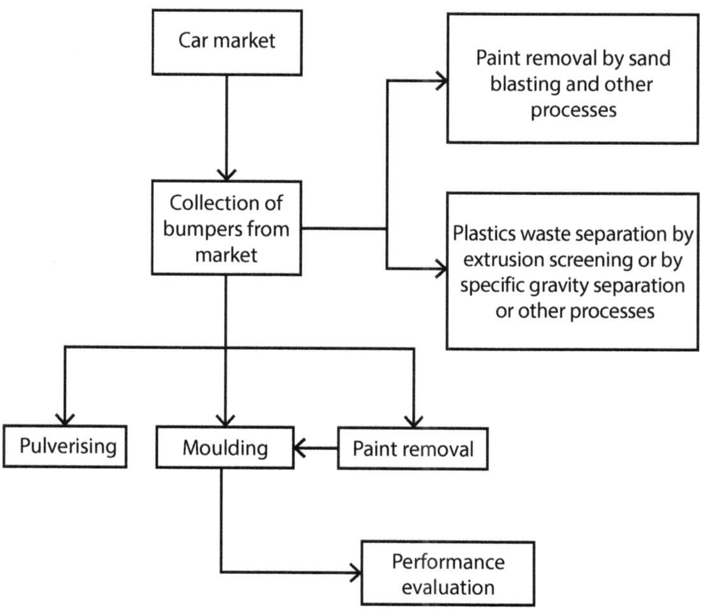

Figure 10.6 Recycling of automobile bumpers.

mirror and others are recycled. Non-painted ABS parts exhibit good properties and surface appearance. Acrylic painted parts exhibit good properties but are partially rough in appearance. Urethane painted parts have low-impact resistance with poor surface appearance. The application of new technology or adoption of mass production may not always be possible in order to achieve productivity improvements in the processing industry. The most practical approach is to review and redesign the processing method and apply automation and mechanization.

10.7 Pros and Cons

10.7.1 Positive Thinking

In 2007, the analysis of plastics waste in Europe reported [34]:

1. The generation of 24.6 million tonnes of postconsumer plastic waste concentrated in the packaging, construction, automotive, and electrical and electronic equipment sectors.
2. 20% was recycled and 30% was recovered as energy.

10.7.2 Negative Effects

There are negative effects if the recycling process is not performed in a thorough manner. Some of the effects are mentioned below:

1. 50% of the waste is disposed of in landfill sites [34].
2. The new 'Waste Framework Directive' has set a 50% recycling target for household waste. The directive does not stipulate what waste should apply to the municipal waste stream as a whole or individual material fractions within this stream [35].

The established hierarchy is prevention, preparation and reuse, recycling and other recovery, such as energy recovery, and disposal. It allows specific waste streams to depart from the waste hierarchy when justified by life cycle thinking [35].

Plastics waste separated from a MSW plant has a high energy content and is therefore considered useful for cogasification with coal, which has been detailed by Romey [36].

To make the large-scale (200–600 MWe) coprocessing of waste economically feasible, sufficient quantities have to be available on a continuous basis for a prolonged period of time and the energy content should not be too low. A substantial amount of plastic waste can be collected through the separation of MSW. Approximately 210 million tonnes of MSW are produced annually in the USA and the same amount again in Western Europe, of which the majority (70–90%) is landfilled [37]. The plastics component of MSW makes up, on average, 8.5 wt% in the USA and 7 wt% in Western Europe (APME 1994). MSW is sifted through in a sieve drum to remove the light material.

10.8 Research and Case Study

Case studies on plastics waste are one of several acceptable methods for analysing the impacts of environmental contamination and estimating the decreased value of a contaminated or previously contaminated property [20]. The areas of plastics waste involve recycling, reuse and isposal; hence, research on waste plastics is, to a large extent, aimed at investigating how this waste is generated, recycled and reused under different conditions. Plastics waste management is carried out by groups and organizations and the consideration of the impact upon society is important for this development. Plastics waste management is a multidisciplinary process involving areas where case studies are of benefit. Many research questions in plastics waste management are suitable for case study research and the case study methodology is well suited for many types/aspects of plastics waste management. Research into plastics waste is a contemporary development and is complicated by the fact the waste components/processes are difficult to study in isolation.

10.9 Remarks

Plastics waste handling technology and corresponding equipment related to transfer of knowledge appears pertinent, the success depends heavily on right waste management procedures to be tailored towards specific location where the technology is employed. Case studies are a very effective tool, but have limitations, and differ from analytical and controlled empirical studies. Case studies have been criticized for being of less value, impossible to generalize from, being biased by researchers and so on; hence, case studies require methodology in order to review and judge the situation properly.

References

1. Baeyens, J., Brems, A., Dewil, R., Recovery and recycling of post-consumer waste materials—Part 2. Target wastes (glass beverage bottles, plastics, scrap metal and steel cans, end-of-life tyres, batteries and house-hold hazardous waste). *Int. J. Sust. Eng.*, 3, 232–245, 2010.
2. Bajus, M. and Hájeková, E., Thermal cracking of the model seven components mixed plastics into oils/waxes. *Pet. Coal*, 52, 164–172, 2010.
3. Plastics Europe. Plastics: The Facts 2013, Plastic Europe: Association of Plastic Manufacturer, Brussels, 2014.
4. Kumar, S., Bhattacharyya, J.K., Vaidya, A.N., Chakrabarti, T., Devotta, S., Akolkar, A.B., Assessment of the status of municipal solid waste management in metro cities, state capitals, class I cities, and class II towns in India: An insight. *Waste Manag.*, 29, 883, 2009.
5. Council Directive 91/156/EEC of 18 March 1991 amending Directive 75/442/EEC on waste, Journal of Laws. L 078, 26/03/1991 P.32.
6. European Parliament and Council Directive 94/62/EC of 20 December 1994 on packaging and packaging waste, Journal of Laws. L 365, 31/12/1994 P.10.
7. Directive 2004/12/EC of the European Parliament and of the Council of 11 February 2004, amending Directive 94/62/EC on packaging and packaging waste. *Off. J. Eur. Union*, L47, 26–31, 2004.
8. *Thematic Strategy on the Prevention and Recycling of Waste*, European Commission, Brussels, Belgium, 2005, COM(2005) 666 Final.
9. Pollution in India – studies on economy and growth – Project work – Dr, Muralisrinivasan Natamai Subramanian, pp. 25–31, 2017- Unpublished
10. Doorasamy, M., Theoretical Developments in Envionmental Management Accounting and the Role and Importance of MFCA. *Found. Manag.*, 7, 37–52, 2015.

11. Saito, K. and Narita, H., Studies on the dechlorination and oil-production technology of waste plastics. *J. Mater Cycles, Waste Manag.*, 3, 93–98, 2001.
12. McAuley, J., France becomes the first country to ban plastic plates and cutlery. *Washington Post.*, e-paper, September, 2016.
13. Smithers, R., England's shoppers say goodbye to free plastic bags. *Guardian*, e-paper, 2015.
14. Trigg, T., England Cuts Plastic Bag Usage by 6 Billion in 6 Months. *Sci. Am.*, October, 2016.
15. Saeed, L. and Zevenhoven, R., Comparison between two-stage waste combustion with HCl recovery and conventional incineration plants. *Energy Source.*, 24, 41–57, 2002.
16. Hamid, S.H. and Amin, M.B., Lifetime prediction of polymers. *J. Appl. Polym. Sci.*, 55, 1385–1394, 1995.
17. World Bank, *Results-based Financing for Municipal Solid Waste. Urban Development Series Knowledge Papers*, World Bank, Washington, DC, 2014.
18. Ruggiero, J.A., *Sociol. Pract. J. Clin. Appl. Sociol.*, 4, 2, 103, 2002.
19. Yin, R.K., Chapter 1, Getting Started: How to Know Whether and When to Use the Case Study as a Research Method, Case study research and applications : Design and methods, 6th edition, pp. 33–60, SAGE publications, Los Angeles, USA, 2018.
20. Jackson, T.O., Appraisal Standards and Contaminated Property Valuation, *Appraisal J.*, 71, 2, 132, 2003.
21. Penrose, E.T., The growth of the firm—A case study: The Hercules Powder Company. *Bus. Hist. Rev.*, 34, 1, 1, 1960.
22. Prahalad, C. and Hamel, G., *Harv. Bus. Rev.*, 79, 1990.
23. Berg, B.L., *Chapter 2, Designing Qualitative Research. Fourth Edition, Qualitative Research Methods for the Social Sciences*, pp. 15–37, Allyn & Bacon, A Pearson Education Company, USA, 1998.
24. Yin, R.K., Chapter 1, Getting Started How to Know Whether and When to Use the Case Study as a Research Method, Case Study Research and Applications:Design and Methods, Sixth Edition, p31–56, SAGE Publications, Inc., Los Angeles, USA, 2018.
25. Stake, R., *Chapter 4, Data Gathering, The art of case study research.* Sage Publications, Thousand Oaks, CA, pp. 49–68, 1995.
26. McTavish, D.Q. and Loether, H.J., *Social Research: An Evolving Process*, 2nd Edition, Allyn & Bacon, pp. 25–27, Boston, MA, USA, 2002.
27. Snow, D.A. and Anderson, L., *A Case for the Case Study*, J.R. Feagin, A.M. Orum, G. Sjoberg (Eds.), p. 148, University of North Carolina Press, Chapel Hill, NC, USA, 1991.

28. Yin, R.K., Chapter 2, Designing case studies, Getting Started How to Know Whether and When to Use the Case Study as a Research Method, Case Study Research and Applications:Design and Methods, Sixth Edition, pp. 57–1166, SAGE Publications, Inc., Los Angeles, USA, 2018.
29. Berg, B.L., *Qualitative Research Methods for the Social Sciences*, 4th Edition, Allyn & Bacon, Boston, MA, USA, pp. 170–180, 2001.
30. Jackson, T.O., Case Studies Analysis: Environmental Stigma and Monitored Natural Attenuation, *Appraisal J.*, 72, 2, 111, Spring 2004.
31. Knappich, F., Klotz, M., Schlummer, M., Wölling, J., Mäurer, A., Recycling process for carbon fiber reinforced plastics with polyamide 6, polyurethane and epoxy matrix by gentle solvent treatment. *Waste Management*, 85, 73–81, 2019.
32. Riess, M., Ernst, T., Popp, R., Müller, B., Thoma, H., Vierle, O., Wolf, M., van Eldik, R., Analysis of flame retarded polymers and recycling materials. *Chemosphere*, 40, 937, 2000.
33. Mäurer, A. and Schlummer, M., Good as new. Recycling plastics from WEEE and packaging waste. *Waste Manag. World*, 3, 33, 2004.
34. The Compelling Facts about Plastics, *An Analysis of Plastic Production, Demand and Recovery for 2007 in Europe*, Plastics Europe, Brussels, Belgium, 2008.
35. Directive 2008/98/EC, The European Parliament and of the Council on waste and repealing certain Directives, Official Journal of the European Union, p. L 312/3 -L 312/28, 2008.
36. Romey, I F.W., "JOULE III programme. Clean coal technology R & D. Advanced combustion and gasification of fuel blends. Volume II. Fuel blends and alkali diagnostics." CEC – Technical Report, 1999.
37. Wallman, P.H., Thorsness, C.B., Winter, J.D., Hydrogen production from wastes. *Energy*, 23, 271, 1998.

11
Present Trends

In the standard history of plastics waste, plastic items being sent to landfills was a common nuisance due to the non-biodegradable nature of plastics. However, over time things got pretty complicated with factories springing up in cities and industrial towns, creating an industrialized society susceptible to the ravages of pollution. However, by then society surrendered to the problem, claiming that it was all too much for the common man to handle. For a while this worked alright by curtailing noxious uses of land. However, plastics waste disposal has become a societal issue of tremendous proportions and an environmental burden. It is one of the most pressing sustainability issues humanity faces. The cleanup of abandoned plastics waste is one issue that already affects virtually every sector of business [1].

11.1 Economic Issues

On a global scale, reuse of plastics waste is showing limited signs of decoupling from economic growth. Plastic materials and product demand are overwhelmingly centered around an economic and technical assessment with limited attention to recycled material and recycled product demand. Every recycling strategy attempts to address the information gap by demonstrating plastics waste can be used to ensure that environmental concerns are significantly integrated into an industrial strategic management approach.

Muralisrinivasan Natamai Subramanian (ed.) Plastics Waste Management, 2nd Edition (219–226)

11.2 Industry and Society

Plastics waste need not prove a cause and effect relationship between the event and the damage. To properly manage the plastics raw material cycles, in principle there needs to be an understanding of the pattern of plastics processing and product flow in industry and society and the transformation it can undergo. It is difficult for individual industries to determine a course of action.

Industry and society both have a metabolism designed to take up, use and dispose of plastic products. Plastics processing often generates waste and consumers use a product and throw it into the environment or dump as waste. A long-term strategy for plastics waste disposal must be based on the following two principles [2]:

1. Shortening product manufacturing by eliminating as many intermediate steps as possible.
2. Improving the utilization of waste.

11.3 Landfilling

As disposal costs increase, the reuse of previously discarded waste from one industry by another is becoming more common. During processing, significant amounts of product end up in recycling bins and dumpsters, which is problematic.

Landfill disposal of plastic waste presents several disadvantages:

- Substantial deprivation of valuable resources from a sustainable point of view.
- Plastic waste is photodegradable (i.e., are decomposed by the sunlight) into very small particles that can be transferred between ecosystems.

Most waste is disposed of in landfill sites, which costs virtually nothing. Landfills have many environmental effects due to:

- Low level of monitoring
- High cost of waste disposal services

Plastics deposition in landfills is not a desirable solution.

11.4 Effect of Single-Use Plastic Products

The increasing number of single-use plastic products in the world is an environmental problem. With plastics waste management it may be possible to prohibit stockpiling of plastics waste with the development of recycling technology for the production of recycled granular material from plastics waste. Plastics should no longer be considered as a pollutant or as cheap landfilling material but should become environmentally friendly materials with utilitarian properties. The main difficulties associated with the recovery and recycling of used plastic products are due to the contaminants present along with it. Due to fusible material, plastics waste can be reprocessed in simple processes like extrusion.

Denmark's Environmental Protection Agency concluded that the plastic grocery bags made up of low density polyethylene are more environmentally friendly than paper, biodegradable plastics and cotton alternatives. Plastic grocery bags made up of low density polyethylene are more environmentally friend than biodegradable materials. However, the dynamic increase of plastics waste, especially in the form of single-use products, has become a serious environmental problem. The low densities of plastic products, such as bottles, shrink wrap, blister packs, etc., has resulted in their having an almost infinite life as landfill. Some of the plastic packaging materials present a hazard to marine life.

11.5 Effect on Food Packaging

Customer satisfaction is key to boosting repeat business, customer loyalty, referrals and, as a consequence, increasing long-term profitability. Food markets have suffered due to the ban on plastic film and packaging. Moreover, the daily market will see a rise in the occurrence of degradation of food materials due to storage without plastic packing. Long distance shifting of product also faces problems from non-compatible packing. Since plastic materials are a barrier to weathering conditions, fungus, bacteria, etc., due to the expense involved in preserving food, large-scale markets will suffer.

Due to the reduction or ban on plastics, old plastics that were being used without any recycling do not have a consumer market.

Because of recycling, the sale of these items changes with the ban. Hence, more of these items will accumulate and will not be able to be used in applications due to not having the additives required. Therefore, the increase in cost is diverted even to medical application products which require sterilization. During sterilization, since they are recycled materials, there may be the possibility of internal cracks or voids in the products or deterioration during sterilization. What is not stored is not managed, and therefore much is spent on preserving the food materials for later use.

11.6 Recycling Status

As it moves towards the future, recycling management is currently in a critical phase due lack of suitable treatment and the increasing amount of waste generated not only from industries but also from consumers [3–5]. Issues relating to a potential waste process at the industrial level are as follows [6, 7]:

- Lack and poor quality of signs and instructions
- Inefficiencies in handling and internal logistics
- Lack and underused bins for waste
- Poor quality of raw material and products
- Process equipment inefficiency
- Recycling inefficiency
- Accumulation of wastes.

Because current plastics consumption has risen, even though substantial recycling technology has been provided, plastics waste has not been reduced up to now. Equally responsible are the market inflexibility and the ensuing policy recommendations. Many industries are already recycling plastics that are below the levels of markets and sales.

11.7 Present Research and Shortcomings

Conducting plastics recycling experiments and reporting the key data have been critical aspects at industrial levels. However, by using laboratory level research around globally relevant questions related to recycling plastics, in connection with plastics processing industries, it is only possible to translate research into commercial processes. During the research stage in the laboratory, the present trend is to translate commercial process results in order to capture different environmental problems.

- The experiments at laboratory levels show the plastics can be mixed and the soil properties could be improved. However, there is no real test conducted in the field due to fear of mixing plastics in the soil, change the entire portion as polluted. All laboratory levels have to bring in the field level with an analysis of their lifecycle with the experiments. However, the lifecycle analysis will take long duration, at that moment the waste generation may go up.
- To reduce environmental burdens, biodegradable polymers are promoted as an environmental solution to plastics pollution. However, their degradation in managed and unmanaged environments is not well known.

The amount of waste increases due to human activities; therefore, a problem of its storage appears. Clearly, many regulations to protect the environment are badly needed to protect society and future generations from the immense damage that the free market system has not been able to take into account. It has begun to circle around the many and complex interdependencies between regulations and industry's innovative behavior [8]. Global production of municipal waste is estimated at 1.2–1.6 billion tons [9].

11.8 Population Growth and Waste

Due to the growth trends of human population and urbanization, there has been a significant increase in the generation of waste worldwide [10]. Waste creates economic and environmental challenges [11].

There has been an increase in virgin resin production at the same time as the promotion of recycling. Postconsumer recyclables are of rather low value in the market place, competing with mass-produced and low-cost virgin materials; therefore, the manufacture of virgin plastics has considerably outstripped plastic recycling.

Environmental regulation has been steadily intensifying to protect the environment. Liberalization of plastic products and their applications has given rise to a new sector of specific regulations, while the emergence of plastics waste has created additional regulatory activities.

Identifying household preference toward waste management service components is an important consideration when successfully introducing waste separation policies [12].

11.9 Remarks

The durability of plastics has been considered to be a positive attribute, but it is now commonly perceived by society as negative. Waste plastics disposal is a complex process and requires active participation by industry, the public and governments. The recycling and reprocessing of waste plastics is driven by the awareness and concern of the detrimental environmental impact of discarded plastics waste. It is an important issue due to the need for resource conservation, and costly and limited energy and feedstock availability. Discarded polymers represent a colossal waste of the energy which is locked within them and if they are not recycled it results in a greater strain on already limited natural resources/feedstock. Polymer recycling is related to the national economy due to hydrocarbon feedstock. The high price of virgin polymers and cheap labour are the driving forces for the recycling of waste plastics.

References

1. Krausmann, F., Schandl, H., Eisenmenger, N., Giljum, S., Jackson, T., Material flow accounting: Measuring global material use for sustainable development. *Ann. Rev. Environ. Resour.*, 42, 647–675, 2017.
2. Ayers, R.U., Industrial metabolism, in: *Technology and Environment*, J.H. Ausubel and H.E. Sladovich (Eds.), pp. 23–49, National Academy Press, Washington, 1989.

3. Mondelli, G., Giacheti, H.L., Boscov, M.E.G., Elis, V.R., Hamada, J., Geoenvironmental site investigation using different techniques in a municipal solid waste disposal site in Brazil. *Environ. Geol.*, 52, 871–887, 2007.
4. Rathi, S., Alternative approaches for better municipal solid waste management in Mumbai, India. *J. Waste Manag.*, 26, 10, 1192–1200, 2006.
5. Li'aoa, W., Ting'quana, P., Chuana, H., Huib, Y., Management of municipal solid waste in the Three Gorges region. *Waste Manag.*, 29, 7, 2203–2208, 2009.
6. Coghlan, D. and Brannick, T., *Doing Action Research in Your Own Organization*, second ed., SAGE Publications, London, UK, 2005.
7. Curran, T. and Williams, I.D., A zero waste vision for industrial networks in Europe. *J. Hazard. Mater.*, 207–208, 3–7, 2012.
8. Bratsev, A.N., Kumkova, I.I., Kuznetsov, V.A., Popov, V.E., Shtengel' S.V., Ufimtsev, A.A., Air plasma gasification of RDF as a prospective method for reduction of carbon dioxide emission. *IOP Conf. Series: Mater. Sci. Eng.*, 19, 012004, pp. 1–7, 2011.
9. Borkiewicz, J., Proekologiczne technologie utylizacji odpadów. *Inżynieria Ekologiczna*, 6, 159–174, 4, 2002.
10. Nizami, A.S., Rehan, M., Ismail, I.M.I., Almeelbi, T., Ouda, O.K.M., Waste biorefinery in Makkah: A solution to convert waste produced during Hajj and Umrah Seasons into wealth, *15th Scientific Symposium for Hajj, Umrah and Madinah Visit*. Madinah, Saudi Arabia, 2015.
11. Ismail, I.M.I. and Nizami, A.S., Waste-based biorefineries in developing countries: An imperative need of time. *Paper presented at The Canadian Society for Civil Engineering: 14th International Environmental Specialty Conference*. Conference Proceedings, London, Ontario, Canada, June 1–4, 2016.
12. Oosterveer, P. and Spaargaren, G., *Meeting social challenges in developing sustainable environmental infrastructures in East African cities. Social Perspectives on the Sanitation Challenge*, pp. 11–30, Dordrecht, Springer, 2010.

12
Future Trends

Science and technology have combined to make our everyday life comfortable, but has placed a burden on the environment in the following ways:

1. Global warming
2. Depletion of natural resources
3. Atmospheric pollution
4. Increase in solid waste
5. Destruction of the ozone layer

Current regulations aim to preserve the environment via recycling materials. Solid waste, particularly plastics waste, is non-degradable and requires economic disposal techniques. However, there is a growing consensus that plastics waste faces challenging operational environments. Due to increasing demand and the subsequent increase of consumer and industrial waste, the amount and composition of plastics waste needs to be characterized prior to disposal. Countries are adopting policy tools to minimize the use of fossil fuels through renewable energy resources and their development and application, along with improving energy efficiency [1, 2].

12.1 Present Problems

- The global energy demand is increasing each year, and the need to develop alternative energy production systems is critical.
- Plastics waste disposal is uncontrolled and rules and techniques for proper landfill management are ignored.

Muralisrinivasan Natamai Subramanian (ed.) Plastics Waste Management, 2nd Edition (227–236)

Therefore, plastics waste has been highlighted as a significant source of the environmental burden.

- Discarded post-consumer plastic products are one of the materials to be collected and recovered. It is chemically inert, non-biodegradable, and thermally stable due to the presence of additives. This allows for infinite reprocessing recovery/reuse operations. The entire amount of recovered plastic discarded products could be reused for a new plastics processing unit. However, only color-sorted and contamination-free waste is reusable in processing. Moreover, if the collection points and processing unit are a long distance apart, the reuse is further prevented. This leads to an increasing amount of discarded consumer plastics waste.
- A majority of the plastics industries are recycling and many others are looking to burn plastics as a source of heat. Therefore, recycling plastics is more desirable than filling up the land and the oceans with them [3].
- Plastics are recycled several times until the properties totally degrade and are no longer useful for any applications. Material that cannot be used for further reprocessing becomes landfill or is incinerated.

12.2 Incineration in Open Air

Burning waste is sometimes seen by consumers as a convenient way to avoid trips to landfill sites, and the municipal corporation is a serious threat to air quality in some developed and developing countries. Pollutants have been linked to many health concerns, including cancer, disruption of endocrine function, developmental problems, endometriosis, cardiovascular disease, asthma and diabetes. Burning the waste also causes fire hazards, unpleasant odours and chemical fog; in addition, the toxic ash can contaminate water supplies [4]. This burning of waste plastics creates atmospheric pollution. The gas emitted while burning municipal waste containing a mixture of plastics forms high density smoke along with poisonous gases like dioxin. Therefore, these gases replace the oxygen level, and many of areas suffer from lack of oxygen. Due to the non-availability of sufficient amounts of oxygen needed, blood cells struggle, leading to loss of hemoglobin in addition to breathing problems [5].

12.3 Environmental Advantages

As a result of the growing awareness of the need for clean air and water, there has been an increased use of pumps which contain plastic parts, such as the casing, casing cover and impeller, as opposed to metallic pumps. A study by pollution engineers revealed that the advantages of plastic pumps are superior corrosion and abrasion resistance, ease of maintenance and scrubbing systems for odour control for use in, e.g., water recovery projects, neutralisation of acidic or alkaline solutions, and the treatment of a wide range of liquid wastes or toxic fumes to satisfy government regulations [6].

12.4 Plastics Waste - Challenge

Circular economy and waste management will stimulate an evolution in the approach to eliminate landfill [7]; however, the new 'Waste Frame Work Directive' has set a 50% recycling target for household waste. The directive does not stipulate what waste can enter the municipal waste stream as a whole or individual material fractions within this stream [8].

Disposal of waste is the act of abandoning items, i.e., placing the waste item somewhere with no intention of moving it anywhere else. Waste disposal is, or should be, a managed activity requiring:

the selection of an appropriate site, the design and engineering of the site and ensuring the processes employed are appropriate, proper management of the entire operation, assessment of the likely consequences and monitoring to confirm acceptability.

Increasing concern regarding the impact of chemicals on health and the environment may have significant impact on the use and disposal of plastics that, in turn, may translate into strict legislation. Environmental regulations require the change of design, manufacture, disposal and use of plastic products in many countries.

12.5 Environmental and Social Problems – Prevention

Plastics waste is becoming an environmental and social problem. Industry can avoid the increasing cost of waste disposal, and potential financial difficulty, by not allowing it to accumulate, which would also benefit the environment. At the same time, in terms of the consumer point-of-view, it is the responsibility of industry to dispose of plastics waste properly. Hence, the plastics industry can incorporate the use of recycled plastics and improve its chances of surviving by:

- Reduce/avoid plastics waste generation by using an efficient manufacturing process.
- Recycle plastics waste by investigating new areas of product application.
- Avoid landfill and the burning of plastics waste by considering environmental and social factors.
- Material recycling does not automatically divert waste from disposal.
- Material that is produced must eventually be disposed of; recycling cannot alter that eventual fate.
- Recycling only reduces disposal if and to the extent that it displaces primary production.

If recycling fails to prevent primary production, for instance, by displacing other secondary materials or by expanding markets, it also will not prevent disposal—it will merely delay that fate.

There can be little doubt that plastics waste faces problems whether it is recycled or disposed of in nature. These problems are, however, are industrial and consumer specific. Consumers are dominated by the use of plastic products, in particular the lack of industrial competitiveness (Figure 12.1).

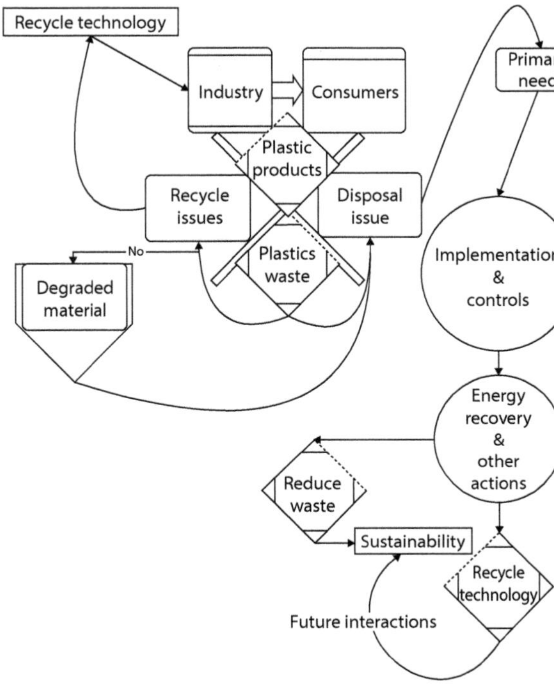

Figure 12.1 Sustainability – plastics waste.

12.6 Reasons – Waste Accumulation

- Opportunities for manufacturing units to promote their own interests instead of the interests of recycling plastics.
- Environmental regulations have begun to become more market oriented in recent years. Regulations are not being steadily intensified in the area of reducing plastics waste accumulation.
- Eco-tax and the implementation recycle management system in other than plastics industries provide strong incentives for internalizing environmental strategies, while at the same time plastics industries retain their entrepreneurial flexibilities.
- Analyzing plastics waste opens the door to complex methodological issues, particularly in terms of sources. By definition, focus on forms of behavior that tend to happen behind closed doors and leave little formal trace.

12.7 Ecological Issues

Each of these ecological crises significantly contributes to a list of organizational problems:

- Ever-increasing regulation on businesses
- Great risks of business project difficulties
- Longer delays in business project planning
- Potentially damaging publicity from technological failure
- The skyrocketing costs of rectifying environmental errors

The pervasiveness of plastics waste is caused in a large part by business-related activity. It has been one of the most important reasons that plastics waste seems to be just the first of many decades in which business needs to pay significant attention to the environment.

12.8 Facts about Bioplastics

Bioplastics represent a wide range of materials that can be produced from renewable sources, petroleum or a mixture of both [9]. They may be bio-based and/or biodegradable. The production of bioplastics is expected to grow in the future.

Bioplastics are degradable. Industries have devoted their attention towards bioplastics due to environmental problems related to non-biodegradable plastics [10, 11].

Bioplastics are not always biodegradable and not always biobased. Bioplastics cannot be considered as a general solution to the litter problem. Biological degradation is very slow and depends on environmental conditions, such as microorganisms, temperature and humidity. Biodegradable plastics may even add to the litter problem due to limitations including the time required for degradation, the use of enzymes, climate change and so on [12].

Analytical considerations for bioplastics waste, in terms of an environmental point-of-view, include:

- Plastics processing should result in optimal physical properties such thermal, mechanical and so on.

- Chemical composition of the plastics material.
- Extraction and migration of chemicals with respect to the environment.

Consideration of the above factors prevents the generation of plastics waste and at the same time helps in the preparation and reuse of plastics waste in addition to recycling and energy recovery. In the sea and on the seashore, it is necessary to ensure the ability of waste plastics to photodegrade and/or biodegrade [13].

12.9 Future Requirements

The plastics industry is primarily concerned with processing and polymer manufacturing with the consumption of non-renewable resources. It is therefore important that the polymer manufacturing industry pursues high environmental standards in its policies and practices, both for reasons of sustainability and the benefit of communities affected by emissions such as harmful gas from polyvinylchloride manufacturing. Within each of the polymer manufacturing major operations, a number of different processes can be identified, such as polymerization, polymer stripping, polymer storage, etc.

Due to increased plastics production and consumption, present trends in plastics waste is one of the main subjects.

- Economies have state-of-the-art waste technology treatments such as incineration, anaerobic digestion, and modern recycling units.
- Through continuous improvement, sustainability can be infused in any plastics production process [14]. There are numerous techniques to eliminate non-value adding production activities [15].
- For use in primary applications, the recycling of plastics waste must be engineered in a clean and segregated closed-loop process [16]. Even packaging material such as film wrap can be collected in bulk from industrial waste; therefore, packaging waste can be economically recycled [17].

- The disposal of unusable plastics wastes requires defined processes, which allow the plastics waste stream to be driven by life cycle thinking.
- Many plastics have a higher calorific value than coal, therefore they can be logically mixed with fossil fuels or used individually in a furnace, which avoids the danger of toxic pollution as a result of not burning waste in open air.
- Reusing plastics is the required course of action, with the additional benefit of reducing emissions associated with plastic production [18].
- Before abandoning incineration technologies, regulators and the market presume that an incumbent industry which invests in radical innovation increases the chances that the new technology will displace the existing waste as standard product.

12.10 Remarks

The recent emphasis on plastics waste and environmental problems among economists has brought into focus the inadequacy of recycling and the economy as sole indicators of well-being. Plastics waste management has been developed for the purpose of international comparisons. A logical extension of plastics waste management research seems to be the design of multidimensional measures which summarize the inequalities with respect to different attributes of recycling or reuse or recovery options.

Two important aspects concerning waste plastics are:

- Multidimensional inequality in the spirit of the environment on solid waste management.
- Merits of perception on plastic materials.

The theory of regulation concerns how to regulate plastics waste with several dimensions of recycling. For instance, it would be useful to know how to regulate

- Recycling with information about marginal costs of each recycled products.

- Industries with information about recycling cost and demand conditions.

The differences in beliefs about solid waste can be completely explained by differences in information. Essentially, the scenario is that solid waste begins with consumers, households and industries and the condition of the wastes last received. If later beliefs differ, it must be due to the fact that waste has been received from distinctly different sources.

References

1. Jebli, M.B., Youssef, S.B., Ozturk, I., Testing environmental Kuznets curve hypothesis: The role of renewable and non-renewable energy consumption and trade in OECD countries. *Ecol. Indic.*, 60, 824–831, 2016.
2. López-Menéndez, A.J., Pérez, R., Moreno, B., Environmental costs and renewable energy: Re-visiting the Environmental Kuznets Curve. *J. Environ. Manag.*, 145, 368–373, 2014.
3. Nuspliger, E.S., Vredeveld, R., Martin, R., Plastics in the Environment. *Am. Biol. Teach.*, 35, 4, 230, 1973.
4. Adamu, H., Luter, L., Lawan, M.M., Umar, B.A., Chemical Speciation: A Strategic Pathway for Insightful Risk Assessment and Decision Making for Remediation of Toxic Metal Contamination. *Environ. Pollut.*, 2(3), 92–99, 2013.
5. Studies on air pollution with plastics –Dr. Muralisrinivasan Natamai Subramanian – project work –unpublished.
6. Gomes, H.I., Mayes, W.M., Rogerson, M., Stewart, D.I., Burke, I.T., Alkaline residues and the environment: A review of impacts, management practices and opportunities. *J. Clean. Prod.*, 112(4), 3571–3582, 2016.
7. The Compelling Facts about Plastics 2007, *An Analysis of Plastic Production, Demand and Recovery for 2007 in Europe*, Plastics Europe, Brussels, Belgium, 2008.
8. Council Directive 2008/98/EC, *Off. J. Eur. Union*, L 312, 3, 2008.
9. Reddy, M.M., Vivekanandhan, S., Misra, M. *et al.*, Biobased plastics and bionanocomposites: Current status and future opportunities. *Prog. Polym. Sci.*, 38, 1653–1689, 2013.
10. Arrieta, M., Samper, M., Aldas, M., López, J., On the use of PLA-PHB blends for sustainable food packaging applications. *Materials*, 10, 9, 2017.

11. Beltrán, F.R., Lorenzo, V., de la Orden, M.U. *et al.*, Effect of different mechanical recycling processes on the hydrolytic degradation of poly(lactic acid). *Polym. Degrad. Stab.*, 133, 339–348, 2016b.
12. Daniels, N., Recycled PVC: An uncertain future?, *Materials World*, p. 11, January 2016.
13. Scott, G., Photo-biodegradable plastics: Their role in the protection of the environment. *Polym. Degrad. Stab.*, 29, 135, 1990.
14. Aoki, K., Transferring Japanese kaizen activities to overseas plants in China. *Int. J. Opera. Prod. Manag.*, 28, 518–539, 2008.
15. Suárez-Barraza, M.F., Ramis-Pujol, J., Kerbache, L., Thoughts on kaizen and its evolution. *Int. J. Lean Six Sigma*, 2, 288–308, 2011.
16. Harata, H., *Proceedings of Recycle'93*, Davos, Switzerland, 22nd–26th March, 1993, Paper No.15/4.
17. Sadrmohaghegh, C., Scott, G., Setudeh, E., Recycling of Mixed Plastics. *Polym. Plast. Technol. Eng.*, 24, 149, 1985.
18. Al-Salem, S.M., Lettieri, P., Baeyens, J., The valorization of plastic solid waste (PSW) by primary to quaternary routes: From re-use to energy and chemicals. *Prog. Energy Combust. Sci.*, 36, 103–129, 2010.

Index

Plastic processing, 58, 64, 71–72

Also of Interest

Check out these published titles from the same author Available at www.scrivenerpublishing.com

Polymer Drugs
By Muralisrinivasan Subramanian
Forthcoming 2020. ISBN 978-1-119-48000-6

Plastics Waste Management, 2nd Edition
Processing and Disposal
By Muralisrinivasan Subramanian
Published 2019. ISBN 978-1-119-55587-2

Polymer Blends and Composites
By Muralisrinivasan Subramanian
Published 2017. ISBN 978-1-118-11889-4

Plastics Additives and Testing
By Muralisrinivasan Subramanian
Published 2013. ISBN 978-1-118-11890-0

The Basics of Troubleshooting in Plastics Processing
By Muralisrinivasan Subramanian
Published 2011. ISBN 978-0-470-62606-1

Lightning Source UK Ltd.
Milton Keynes UK
UKHW022239251019
352315UK00003B/114/P

9 781119 555872